T0245231

CAMBRIDGE LIBRARY COLLECTION

Books of enduring scholarly value

Technology

The focus of this series is engineering, broadly construed. It covers technological innovation from a range of periods and cultures, but centres on the technological achievements of the industrial era in the West, particularly in the nineteenth century, as understood by their contemporaries. Infrastructure is one major focus, covering the building of railways and canals, bridges and tunnels, land drainage, the laying of submarine cables, and the construction of docks and lighthouses. Other key topics include developments in industrial and manufacturing fields such as mining technology, the production of iron and steel, the use of steam power, and chemical processes such as photography and textile dyes.

Mine Drainage

Stephen Michell's 1881 work covers the full range of engines and steam-pumps available for draining mines in the nineteenth century. An expert on contemporary mining technology, Michell co-authored the essays 'The Best Mining Machinery' and 'The Cornish System of Mine Drainage' prior to writing this comprehensive survey. *Mine Drainage* represents the first attempt to gather in one book information previously located in various journals (and therefore difficult to find), and documentation about engines by their (possibly biased) manufacturers. The book also contains almost 140 illustrations of the diverse pumps and engines discussed. After a short introduction, the material is organised into two main sections, focusing on horizontal and vertical engines. Within those categories it discusses rotary and non-rotary engines, and simple and compound steam-pumps. The book will interest historians of technology, science, engineering, and mining in the Victorian period.

Mine Drainage

Being a Complete and Practical Treatise on Direct-Acting Underground Steam Pumping Machinery

STEPHEN MICHELL

CAMBRIDGE UNIVERSITY PRESS

Cambridge, New York, Melbourne, Madrid, Cape Town, Singapore,
São Paolo, Delhi, Dubai, Tokyo, Mexico City

Published in the United States of America by Cambridge University Press, New York

www.cambridge.org
Information on this title: www.cambridge.org/9781108026659

This edition first published 1881
This digitally printed version 2011

ISBN 978-1-108-02665-9 Paperback

MINE DRAINAGE

A TREATISE ON

UNDERGROUND STEAM-PUMPING MACHINERY

LONDON: PRINTED BY
SPOTTISWOODE AND CO., NEW-STREET SQUARE
AND PARLIAMENT STREET

MINE DRAINAGE

BEING

A COMPLETE AND PRACTICAL TREATISE

ON

DIRECT-ACTING UNDERGROUND STEAM PUMPING MACHINERY

WITH A DESCRIPTION OF A LARGE NUMBER OF THE BEST KNOWN ENGINES, THEIR GENERAL UTILITY AND THE SPECIAL SPHERE OF THEIR ACTION, THE MODE OF THEIR APPLICATION, AND THEIR MERITS COMPARED WITH OTHER FORMS OF PUMPING MACHINERY

BY

STEPHEN MICHELL

JOINT-AUTHOR OF ESSAYS ON 'THE BEST MINING MACHINERY' AND 'THE CORNISH SYSTEM OF MINE DRAINAGE'

LONDON

ROSBY LOCKWOOD AND CO.

7 STATIONERS'-HALL COURT, LUDGATE HILL

1881

PREFACE.

The system of underground pumping by means of direct-acting engines placed near the water is now a most important one, and promises to become more so. The author believes therefore that an account of some of the most prominent engines belonging to it, the principle of their construction, the method of their action, their capacities, prices, working economy, &c., and the comparative value of this system of mine drainage, will be of some service to mining engineers and capitalists. The addresses of the makers of the engines, which are given in every case, will be found a convenience to those who may have to communicate with them. In describing the various engines, he has, as far as possible, endeavoured to give such details as will be interesting both to the engineer and miner; and to deal with the practical results, which more particularly concern the miner and which are the only considerations to which he is likely to give his serious attention. It may be well to state, in order to explain the absence of the names of many first-class makers, whose engines, it may be, are as excellent as any that are noticed, that it formed no part of the author's task to furnish an index to the very numerous manufacturers of the engines comprised in this system of drainage. Such information must be sought in the trade journals and business directories. The examples given include makers of the highest repute, whose names are familiar in the mining districts of the two hemispheres, and the address being given in every case, whoever makes mining his pursuit will observe where he may procure, for any drainage operations he may have to prosecute, engines of the best manufacture, whether it be a diminutive specimen for light, auxiliary work in a 'dip,' 'winze,' or any other place remote from the main engine, a larger engine to assist in unwatering flooded workings, or a machine of dimensions huge enough to contend single-handed with the water which pours in copious streams into the lower parts of deep mines.

When steam was first successfully resorted to for the purpose of pumping water out of mines, the engine was placed at the top of the shaft or pit, and made to operate upon appliances which extended to the bottom, where the water collected. That system continued, for many generations, the only mode of pumping by steam-power. When the depth of the pit became such that manual, horse, or water-power was insufficient to pump the water out, there has been no resource but to adopt the surface engine with its massive masonry and extensive material, involving great cost; which in many instances has been needless, as further exploitation of the mine has determined its unprofitable character. This system of drainage is known as the 'Cornish,' from the early and extensive application of it to the Cornish mines, and the Cornish engine has been the most famous machine ever developed for the unwatering of mines, and in districts remote from the coal-fields it is likely to continue in use for deep and extensive drainage.

The great want for many years has been a smaller, more compact, less costly machine, that should be sufficient to carry the development of mines beyond the very limited extent attainable by hand, horse, or water-power drainage. These means would not permit an experiment to be carried sufficiently far, in many cases, to determine the mineral resources of a locality, and the experimentalists would either be obliged to abandon their undertaking or hazard the outlay of a costly Cornish engine. The necessity of a machine that would serve as a midway appliance between animal- and water-power and a huge surface engine—something that should supersede the latter for experimental mining and all light drainage operations—has been keenly felt for a long period, and it is only within the last few years that the want has been met by the introduction of the various forms of underground engines known as 'steam-pumps.' The principle of employing engines underground to force the water upwards by direct pressure is one which is receiving increasing attention from engineers, and the application during the last few years of the compound system of cylinders has extended the principle to deep and extensive drainage; and in coal mines, where economy of fuel is of far less consequence than in metalliferous mines, the development of this system of drainage has been most surprising. The stimulus which its ready adoption gave to the manufacture of steam-pumps led to the appearance of a large number of engines of various kinds, many of which will be described in the following pages.

To these younger products of the engineer's skill, the surface engine may relegate much of its work, or it may engage them as auxiliaries, for which their much greater lightness of structure, their ready portability, and the celerity with which they may be set to work, make them so well adapted.

Many admirable papers on underground pumping engines have appeared in the journals and transactions of the engineering societies from time to time, but in this scattered state they are obviously only accessible to a very limited number of persons. Moreover, the high price of some of the journals containing these papers, and the trouble and perhaps impossibility of bringing together such of these isolated papers as might be required, would prevent the general reader availing himself of them. Many of these papers are the production of promoters of the engines to which they refer, and who might naturally be supposed to be prepossessed in favour of their own machinery. The author is not aware that any work has been written dealing in a comprehensive manner with this important system of drainage, and he hopes the following pages will help to meet a long-felt want. That they do not supply more information on such important matters to the engineer and miner as the economy or 'duty' of the different machines is regretted, but this is not the author's fault. A thorough trial of the different forms of engines under circumstances fair to each has not been made, and so any remarks on their relative value as pumping agents must be considered as only an approximation to the truth. When a committee composed of independent mine managers and engineers, who have made the subject of mine drainage and pumping machinery a special study, and sufficiently representative of the different mining centres and systems of pumping, has been formed for the purpose of thoroughly investigating the subject and making complete trials, extending over a lengthy period and in several and diverse mining districts, then and not till then will the exact absolute and comparative values of the various engines be satisfactorily established.

In conclusion, the author has much pleasure in thanking Mr. P. R. Björling, M.S.E., of Birmingham, and Mr. James Colman, of Perranarworthal, for their invaluable help. He also acknowledges with thankfulness important aid given by Mr. Robert Hunt, F.R.S., Keeper of the Mining Records, and Dr. Le Neve Foster, for several years Her Majesty's Inspector of Mines for Cornwall, Devon, &c., who have done so much to encourage the application of the principles of science to the operations of mining; and with equal pleasure he thanks Mr. W. Galloway, formerly Assistant-Inspector of Mines for South Wales, Mr. John J. Thomas, Manager of the Tynewydd Colliery, and other gentlemen, who have done much to further the preparation of this work.

TABLE OF CONTENTS.

LIST OF ILLUSTRATIONS.

UNDERGROUND
STEAM-PUMPING MACHINERY.

INTRODUCTORY.

For many generations the Cornish engine has maintained undisputed pre-eminence for the drainage of mines, and in all parts of the civilised world has been found the ponderous structure dipping its enormous trunk low down into the bowels of the earth and lifting to light the fountains of the deep; and so for a century it has pioneered the way for the adventurous miner and enabled him to reach the vast mineral wealth lying deep in the earth's crust. From the moment Watt made the machine a perfect instrument, it has been, practically, without a competitor until the last decade. Unexcelled and even unapproached for permanent and deep and extensive drainage, it was too costly and massive for many drainage operations, and the ingenuity of the engineer has been tasked to produce a machine that would be suitable for those special forms of drainage, and the application of thought and the growing knowledge of an ever-advancing civilisation, in the last quarter of a century, have left their impress upon our commercial history in the particular branch of mechanics we are considering, in the invention and subsequent development of many forms of pumping machinery. The progress has been characterised by the birth, or rather, the revival of many ideas—for the appliances we now possess are, in a great many instances, merely the practical application of the ideas of our forefathers—for the design of machines in which the motive power should be exerted upon the liquid to be raised in a direct manner, and thus to simplify the construction, and consequently lessen the cost of manufacture, of pumping machinery. The most important additions to our pumping appliances are undoubtedly the very numerous class of Steam-Pumps. It does not appear that steam-pumps suitable for mine drainage were known, at least not in the forms now so familiar, until about thirty years ago. In the Great Exhibition of 1851 there was only one steam-pump exhibited. It had a vertical cylinder and a fly-wheel. The non-rotary horizontal pumps first made their appearance about ten years later. There are several varieties of pumps, and we shall separate them into two great groups—namely, Horizontal pumps and Vertical pumps. Each group may be divided into Rotary and Non-Rotary.

The Horizontal group, which is the most important, may be subdivided into Simple Steam-Pumps and Compound Steam-Pumps: the first embrace by far the greatest number of machines, and which are very similar in construction and mode of action, the main point of difference being in the steam-valve gear. These are the most important class of Simple Pumps, and many thousands are in use in this and other countries, not only in mines, but in tanneries, chemical-works, waterworks, &c. &c. The second are a more economical class than the single-cylinder engines, having been introduced to economise the consumption of fuel.

Vertical Steam-Pumps are very useful for all or most of the purposes for which the horizontal pumps are more generally used, and for pumping under certain circumstances they are preferable. Some of the varieties of these pumps are merely the ordinary horizontal steam-pumps arranged to work vertically.

Although not suitable for all pumping operations—no machines are—steam-pumps occupy nevertheless a high position amongst modern pumping appliances, and have rendered important services in the prosecution of mining, particularly in experimental operations in the opening up of new mines, and in sudden emergencies such as the breakage of other pumping machinery, and the flooding of underground workings. It is well known what valuable services these pumps performed at Tynewydd when the water broke in from an adjoining colliery, flooded the bottom workings, and imprisoned several men.

Before proceeding to discuss the qualities of steam-pumps, it will be necessary to propose the question—What is a steam-pump? An answer is not easily found; the term is so elastic and so loosely applied, and may be made to comprehend so much. It may be observed that machines of precisely similar character bear different names. One maker will call his machine, say, the 'Victoria' Steam-Pump, another will give a similar machine the more pretentious designation of, say, the 'Albert' Pumping Engine. We are afraid that a satisfactory answer cannot be given to the question, but engineers appear to apply the term to any compact machine, vertical or horizontal, for pumping water (or other liquid) in which the power of the steam is applied directly to a plunger or piston (one or more) moving in a water-cylinder or water-cylinders, and the steam and water parts are either on a common bed-plate or foundation, or are otherwise so immediately and closely connected as to form, as it were, one machine. The term is not entirely restricted to single-cylinder engines; but in the case of a combination of steam-cylinders, the engines are known as Compound Steam-Pumps, in contradistinction to the Simple Pumps. By many engineers the term appears only to be applied to the single-cylinder engines; the compound pumps, from their greater size and importance, taking rank with other and more extensive forms of machinery, and for which the term steam-pump is an insignificant designation.

Any steam-pumping machinery may be called a steam-pump. The Cornish Engine, with its Pitwork, may be called a steam-pump: that is, the pump or pumps in the shaft are moved by the agency of steam. But, if we accept the foregoing definition, the Cornish Engine and Pitwork cannot be thus designated—first, because it is not direct-acting;

secondly, because it is not compact; and, thirdly, because the steam and water parts are not so closely connected as to form one machine. The Bull Engine and Pitwork, although direct-acting, cannot be considered a steam-pump in the sense of the foregoing definition. With this, as with the Cornish Engine, the pumping part is an adjunct, but in steam-pumps the pumping part is a component of the machine. True, the Cornish and Bull Engines are not complete without the pumps, but it is not an absolute necessity that either of these engines should be used with the pumps. Other forms of engines, and even water and wind power, may be and are applied to work the pumps.

An order to a maker for a steam-pump would mean a machine with the steam and water parts complete, without, of course, the pipes, but if a Cornish Pumping Engine be ordered, only the steam-engine would be produced; the pumping parts would be quite a distinct matter, and might be had from another maker. The Pulsometer—than which no form of pumping machinery can be more literally a steam-pump in the principle of its action—is quite a different machine from the older and more extensively useful classes of pumps which form the subject of the present and following articles, and has only a very limited application.

The earliest steam-pump particularly applicable to the drainage of mines we are acquainted with was that invented by Mr. William Elliot Carrett, of Leeds, and registered August 31, 1850. This was a vertical engine having a lift and force-pump combined, and a fly-wheel.*

* 'A steam-pump, combining a high-pressure engine, and an improved suction and force-pump, designed and constructed for filling low or high-pressure locomotive, stationary, or marine steam-boilers, and for fetching or forcing water any distance or height; it may, by disconnecting the pump, be used as a steam-engine for driving portable machinery for engineering works, the household or farmyard, for working hydraulic presses, water-cranes, &c. . . . . It is constructed to fetch or force water any required distance *in one continuous stream*, without shock or injury to the pipes or machinery, and at an effective velocity. Mr. William Elliot Carrett, 13 Rockingham Street, Leeds, inventor and patentee.'—*Official, Descriptive and Illustrated Catalogue of the Great Exhibition of* 1851. Vol. i. p. 214.

About three years later Mr. John Cameron, of Salford, invented his vertical steam-pump. In the United States the 'Blake' pump made its appearance in the year 1862. An illustration of the first pump made represents a very simple and unpretentious little machine. This appears to be the first pump of the kind brought to a practicable issue, and it was soon followed by many others of the same class, both in this country and the United States. About the year 1865 Mr. Gilbert Lewis, of Manchester, took out a patent for a pump of the same class, and two years later—in 1867—the pump invented by Mr. A. S. Cameron, of the United States, was introduced into this country, and is now well known as the 'Special' steam-pump of Messrs. Tangye Brothers and Holman. The 'Universal,' another very well-known pump, is also of American origin, being the invention of Messrs. Ezra Cope and James Riley Maxwell, of Cincinnati: it was patented in England, September 15, 1868. These pioneers have had a numerous following in such kindred pumps as the 'Selden,' 'Imperial,' 'Standard,' 'Excelsior,' 'Challenge,' 'Victoria,' 'Lancashire,' 'Caledonia,' and 'Colebrook's' pumps—some of which, however, were quickly abandoned. Besides these, many vertical pumps have made their appearance since the introduction of Mr. John Cameron's well-known pump, such as the 'Simplex' and the pumps of Messrs. Frank Pearn, Wells and Co.; Hulme and Lund; May and Mountain; Carrick and Wardale; and William Turner. More recently the more economical principle of compound cylinders has been applied to the non-rotary horizontal type of pumps. The best-known engines of this class are those of Messrs. Davey, Parker and Weston, Sturgeon, Cherry, Field and Cotton, Cope and Maxwell, and Tangye Brothers and Holman. The simplest, and for our purpose, perhaps, the best mode of classifying steam-pumps will be to separate them primarily into groups according to the direction or position of the cylinder. Such an arrangement will give us the two following groups, a subdivision of each being necessary for rotary and non-rotary engines.

I. Horizontal Engines.

A. *Rotary.*	B. *Non-Rotary.*
A. Simple Steam-Pumps.	A. Simple Steam-Pumps.
B. Compound Steam-Pumps.	B. Compound Steam-Pumps.

II. Vertical Engines.

A. *Rotary.*	B. *Non-Rotary.*
A. Simple Steam-Pumps.	A. Simple Steam-Pumps.

A large variety of engines is represented by the different classes given above, as will be observed from the following detailed arrangement, and which may be considered a complete classification of all the known forms of underground steam-pumping engines. It will give the reader at a glance an idea of the activity that has been displayed in the production of divers forms of underground pumping engines.

As they will be considered in the order of this classification, the reader will be assisted in referring to the different classes and the engines comprised in each.

CLASSIFICATION OF UNDERGROUND STEAM-PUMPING ENGINES.

I. Horizontal Pumping Engines.

A. Rotary Engines.

A. *Simple Steam-Pumps.*

1. Single steam-cylinder, expansive, non-condensing engines, working double-acting pump by connecting-rod and crank, regulated by fly-wheels.

F. Pearn, Wells & Co's Ram-pump.
W. Turner's ,,
,, ,,
,, Piston-pumps.

2. Two steam-cylinder, expansive, non-condensing

engines, working double-acting pumps by connecting-rod and crank, regulated by a fly-wheel.

W. Turner's Piston-pump.

3. Three steam-cylinder, expansive, non-condensing engines, working double-acting pumps by connecting-rod and crank, regulated by a fly-wheel.

W. Turner's	Ram-pump.
John Warner & Sons'	„

B. *Compound Steam-Pumps.*

1. Compound steam-cylinder, expansive, condensing engines, working double-acting pumps by connecting-rod and crank, regulated by fly-wheels.

W. Turner's Piston-pump.

B. Non-Rotary Engines.

A. *Simple Steam-Pumps.*

1. Single steam-cylinder, non-expansive, non-condensing engines, working double-acting piston or ram-pump.

The 'Special'	pump.	
The 'Universal'	„	(short stroke)
„	„	(long stroke)
Colebrook's	„	
Davey's 'Differential'	„	
The 'Blake'	„	
The 'Caledonia'	„	
The 'Selden'	„	
The 'Imperial'	„	
The 'Standard'	„	
Turner's Tappet	„	

2. Single steam-cylinder, expansive, non-condensing engines, working double-acting pump.

Davey's Patent 'Differential' Engine.
The 'Self-Governing' Engine.
W. Turner's 'Tappet Expansion' Engine.

3. Single steam-cylinder, expansive, condensing engines, working double-acting pump.

Davey's 'Differential' Engine.
Parker and Weston's „

B. *Compound Steam-Pumps.*

1. Compound steam-cylinder, expansive, condensing engines, working double-acting pump.

Davey's 'Differential' Compound Engine.
The 'Self Governing' „
Cherry's „
Sturgeon's „
Parker and Weston's „
Field and Cotton's „
Walker's „
Tangye's „

II. Vertical Pumping Engines.

A. Rotary Engines.

A. *Simple Steam-Pumps.*

1. Single steam-cylinder, expansive, non-condensing engines, working pump by connecting-rod and crank, regulated by a fly-wheel.

a. One steam-cylinder and one single-acting ram.

John Cameron's pump.
Ashworth Brothers' „
The 'Manchester' „
The 'Salford' „
Hulme and Lund's „

b. One steam-cylinder and two single-acting rams.

John Cameron's pump.

c. Two steam-cylinders and two single-acting rams.

John Cameron's pump.
Ashworth Brothers' „
The 'Manchester' „
Hulme and Lund's „

d. One steam-cylinder and one double-acting ram.

The 'Simplex' pump.
Pearn, Wells and Co's „

e. Two steam-cylinders and two double-acting rams.

The 'Simplex' pump.
Pearn, Wells and Co's „

f. One steam-cylinder and one double-acting piston and D slide-valve pump

Carrick and Wardale's pump.
Turner's „

g. One steam-cylinder and one single-acting ram and single-acting piston.

May and Mountain's pump.

B. NON-ROTARY ENGINES.

A. *Simple Steam-Pumps.*

1. Single steam-cylinder, non-expansive, non-condensing engines, working double-acting pump.

The 'Special' pump.
The 'Universal' „
The 'Imperial' „

I.

HORIZONTAL PUMPING ENGINES.

A. *ROTARY HORIZONTAL ENGINES.*

THE ROTARY PRINCIPLE has not been readily adopted in mines and collieries for horizontal engines. More room is required, and as a rule the engines have more parts, and are more costly than the non-rotary engines. They are heavier, and consequently are less portable, and not so easily fixed. But none of these can be deemed vital objections to their use for general underground pumping of the light character performed by simple steam-pumps. They possess such important advantages over non-rotary pumps as greater economy of working and certainty of action. By using a heavy fly-wheel or wheels, great regularity in the strokes is obtained, and erratic movements are impossible; the piston may work to within a quarter of an inch of the covers without any risk of knocking them out, and it is almost impossible for the engine to become 'centred.' In pumps of the non-rotary class it is customary to leave a clearance of from 1 to 3 inches, to avoid injury to the covers. In the remarks on Vertical Steam-Pumps the respective values of rotary and non-rotary engines are referred to at greater length.

Notwithstanding the many admirable properties which these pumps possess, mining engineers generally do not favour the application of the rotary principle to horizontal underground pumping engines, consequently these pumps are not very largely used for underground drainage, and their position amongst pumping engines is therefore not a prominent one. These circumstances will be a sufficient justification for not considering these machines with as much detail as those of the non-rotary class.

There are two classes of these engines—namely, simple and compound; but the compound principle is seldom adopted.

A. SIMPLE STEAM-PUMPS.

THESE engines consist of one steam-cylinder with one or two fly-wheels, but generally two. The fly-wheels are fixed behind the water part of the pump, one on each side, and work on the same crank-shaft, and they operate a pair of cranks, one of which is connected by means of a rod to one side of the ram and the other to the other side. The ram works in two cases or working-pumps, thus forming two single-acting pumps, or one double-acting pump. The air-vessel is attached to a delivery-pipe, which is connected to the two delivery-branches. The steam-cylinder in these engines forms a somewhat insignificant factor at one end of the machine, being dwarfed by the fly-wheels. The steam-valves are of the ordinary slide-valve type, and are actuated by eccentrics on the crank-shaft. The pump-valves are either of the kinds known as 'butterfly' and 'wing' valves.

The details are variously arranged, but there are only two main classes, namely, ram-pumps and piston-pumps.

In the engines illustrated by figs. 1 and 2 there is a long ram working into two ram-cases, with two fly-wheels mounted at the back of the pump and two connecting-rods passing from the cranks on the fly-wheel shaft to the ram.

In the engine shown by figs. 3 and 4 it will be observed that the steam-cylinder is placed between two single-acting pumps of the ram type.

Figs. 9 and 10 illustrate a triple-cylinder engine consisting of three pairs of single-acting ram-pumps, or three double-acting pumps of the last type mentioned, each pair of pumps having its own steam-cylinder.

In another variety of the ram-pump the fly-wheels are mounted between the steam-cylinder and the pump: the fly-wheel shaft intersecting the piston and pump-rods, a slot-link is provided to permit of the action of the crank across the rods. The pump consists of a long ram working through two ram-cases and forming two single-acting pumps.

There is still another arrangement of the ram-pump, in which, by the use of a tooth-wheel geared in a pinion on the fly-wheel shaft, the speed of the pump (consisting of three single-acting rams) is reduced below that of the engine, a corresponding increase of power in the pump being the result.

At pages 23 and 24 four varieties of double-acting piston pumps are noticed. In the first variety, fig. 5, the steam-cylinder is fixed upon one end of the bedplate, and the pump upon the other end, the fly-wheels being mounted between the two.

In the next variety a pair of fly-wheels are mounted at one end of the engine, behind the pump, and the steam-cylinder occupies the other end. This engine is shown at fig. 6.

In another variety of the piston type the pump is at one end, the fly-wheel at the other, and the steam-cylinder between the two. Figs. 7 and 8 represent a pair of these engines, with one fly-wheel, and a pair of cranks at right angles to each other.

Sizes, Weights, and Speed.

Sizes.—These engines can be made of any capacity to suit the work to be performed. For engines of equal capacity they occupy more room than the non-rotary engines. The following table gives the dimensions of the foundation-plate of various sizes of engines and the space they require. The actual space required for the engines is greater than the measurements given of the foundation-plates, as the fly-wheels and steam-cylinders generally project beyond the plate. If we add one-third to the lengths given and one-fourth to the widths, we shall approach very nearly the actual area covered by the engines.

Diameter of steam-cylinder	Diameter of pump	Stroke	Dimensions of foundation-plate	Space required
in.	in.	in.	ft. in. ft in.	ft. in. ft. in.
4	2	4	3·2 × 1·1	4·3 × 2·3
4	2½	4	4·6 × 1·8	6·0 × 2·6
4	2½	5	4·6 × 1·8	6·0 × 2·6
6	3	6	4·6 × 1·8	6·0 × 2·6
7	3½	7	5·3 × 2·0	7·6 × 2·6
8	4	8	5·3 × 2·0	7·6 × 2·6
9	4½	9	6·0 × 2·3	9·6 × 3·0
10	5	10	6·0 × 2·3	9·6 × 3·0
12	6	12	8·9 × 3·0	12·0 × 3·0
10	5	15	—	13·0 × 3·0
12	6	18	—	15·0 × 4·0
14	7	21	—	16·0 × 4·0
18	9	24	—	20·0 × 5·0
30	12	28	—	22·0 × 6·0

Weights.—These engines are heavier than non-rotary engines of equal capacity, owing to the heavy fly-wheels. Compared with the last-mentioned pump in the following table, a 12-in. steam-cylinder engine of the non-rotary class, with an 8-in. pump and 12-in. stroke, weighs only 22 cwt., or about ⅓rd of the engine named.

Dia. of steam-cylin.	Dia. of pump	Stroke	Weight	Dia. of steam-cylin.	Dia. of pump	Stroke	Weight
in.	in.	in.	tons cwt. qrs. lbs.	in.	in.	in.	tons cwt. qrs. lbs.
4	2	4	0 2 0 0	8	4	8	0 14 0 0
4	2½	4	0 7 0 0	9	4½	9	1 2 0 0
4	2½	5	0 7 1 0	10	5	10	1 16 0 0
6	3	6	0 9 3 0	12	6	12	3 0 0 0
7	3½	7	0 12 0 0	—	—	—	—

Speed.—The following tables give the speed and other particulars, for moderate lifts, of Mr. Turner's Double Ram-Engines (see pp. 19, 22).

Double Rams with Short Stroke.

Dia. of steam cyl.	Dia. of pump	Stroke	Suction	Dely.	Steam	Exhaust	Feet per minute	No. of revol.	Gallons per hour
in.	in.	in.	in.	in.	in.	in.			
6	3	6	3	2½	1¼	1½	85	85	1300
7	3½	7	3½	3	1¼	1½	93	80	1900
8	4	8	4	3	1½	2	100	75	2700
9	4½	9	4½	3½	1¾	2¼	105	70	3600
10	5	10	5	4	2	2½	108	65	4500
12	6	12	6	5	2½	3	120	60	7200

Double Rams with Long Stroke.

Dia. of steam cyl.	Dia. of pump	Stroke	Suction	Dely.	Steam	Exhaust	Feet per minute	No. of revol.	Gallons per hour
in.	in.	in.	in.	in.	in.	in.			
8	4	12	4	3	1½	2	120	60	3200
10	5	15	5	4	2	2½	125	50	5200
12	6	18	6	5	2½	3	130	43	7800
14	7	21	6	5½	3	3½	140	41	11500
16	8	24	8	7	3	4	150	37	16000
18	9	24	8	7	3½	4	150	37	20300
20	10	28	9	8	3½	4½	160	34	27000
24	12	28	10	9	4	5	160	34	38400

Capabilities.

These pumps may be employed for general underground drainage of the light character performed by other steam-pumps. They are obviously not suitable for pumping out shafts, as they cannot easily be swung upon a movable frame. They occupy more room than other forms of pumps, and being comparatively very heavy it would not be an easy task to move a large engine, as the water decreases. The fly-wheels are usually massive, and would be very much in the way in a shaft.

For use at the bottom of a shaft, or in a sump or gallery, where there is sufficient space, it will be found to be very useful and reliable, and less extravagant of fuel than simple non-rotary pumps.

The engines can be made to force to any height attainable by other steam-pumps. Mr. W. Turner, of Salford, a maker of a large variety of direct-acting pumping engines, has made for Zabreze and Königshütte, in Germany, engines of this class having 47-inch and 45-inch steam-cylinders, with 15-inch and 9-inch pumps respectively, and 3-feet stroke, working against a head of about 900 feet.

Cost.

Rotary are more costly than non-rotary pumps. The difference in small sizes is about 15 per cent. and in large sizes about 25 per cent. The reader may compare the prices given below with those in the article on non-rotary pumps for similar size engines.

If the pumps be fitted with a condenser and a blow-through valve, the price will be increased by about 20 per cent.

Dia. of steam cylinder	Length of stroke	Gallons per hour	Price	Dia. of steam cylinder	Length of stroke	Gallons per hour	Price
in.	in.		£	in.	in.		£
4	9	950	23	10	12	8.640	75
5	9	1,500	26	12	12	11.590	88
6	9	2,160	30	12	18	15,360	110
6	9	2,940	35	14	18	19.440	135
7	12	3,480	43	14	24	24,000	165
7	12	4,860	52	16	24	34.560	220
8	12	6,000	62	18	24	46,360	290

FRANK PEARN & Co
PATENTEES & MAKERS
MANCHESTER
HARE

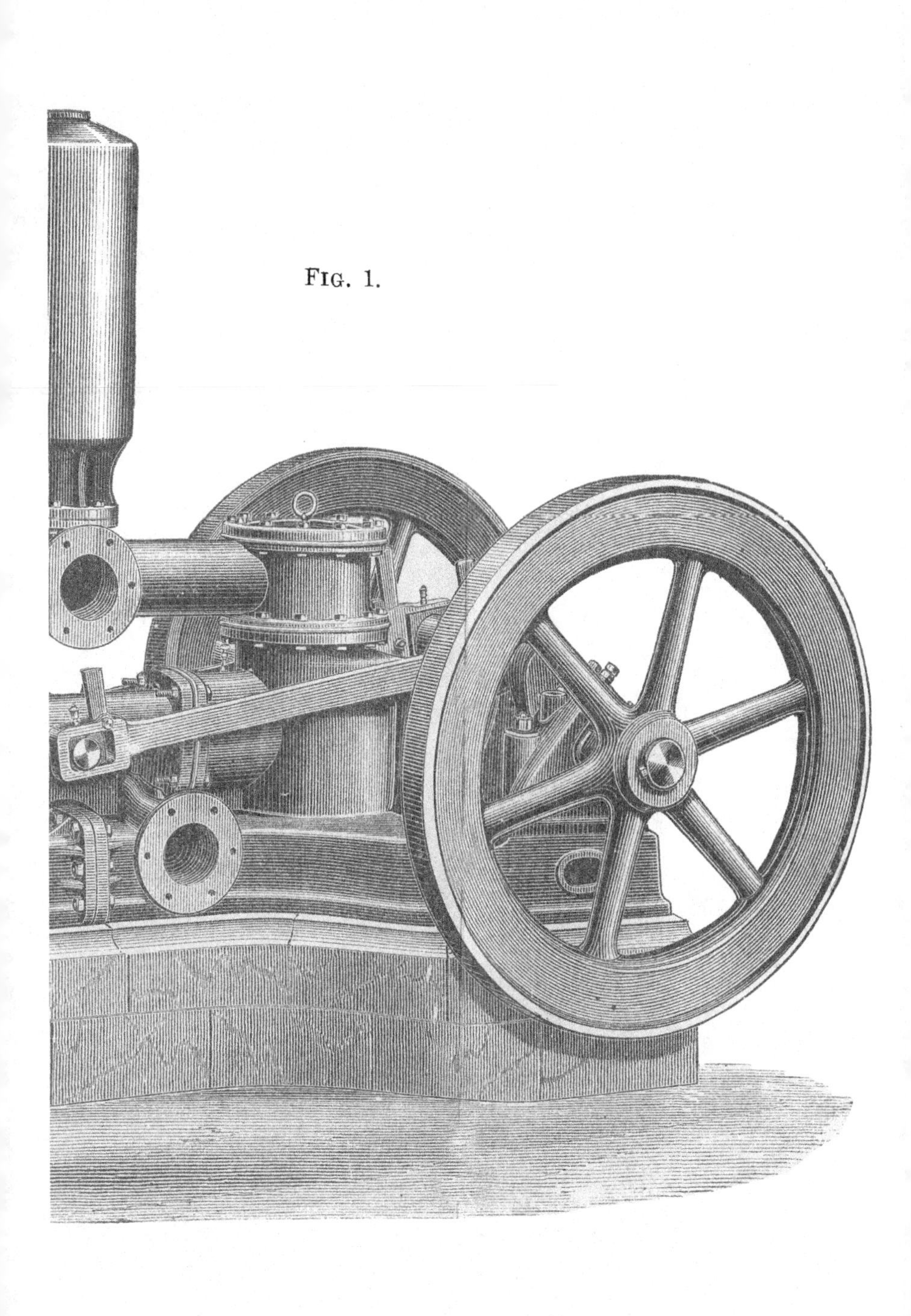

Fig. 1.

1. SINGLE STEAM-CYLINDER, EXPANSIVE, NON-CONDENSING ENGINES, WORKING DOUBLE-ACTING PUMP BY CONNECTING-ROD AND CRANK, REGULATED BY FLY-WHEELS.

Frank Pearn, Wells and Co.'s Rotary, Horizontal, Double-acting Ram-Pump.

It will be observed that between this and the following pump of Mr. W. Turner there is no appreciable difference. The engravings are so self-explanatory, and so much having already being stated concerning the general principles of these engines, it will be needless to give a detailed description of them. The sizes of Messrs. Frank Pearn, Wells and Co.'s engine (fig. 1) are advertised from 4-inch to 12-inch rams, with from 8-inch to 24-inch strokes. The delivery per hour of these two sizes is 3,840 and 46,080 gallons respectively. The address of this firm is West Gorton, Manchester.

Messrs. Hulme and Lund, of Salford, Manchester, are makers of a similar machine.

Turner's Rotary, Horizontal, Double-acting Ram-Pump.

This engine is illustrated by the accompanying elevation, fig. 2. There are two fly-wheels, but only one is shown in order that a better view of the engine may be given. It is made in sizes varying from a 6-inch to a 30-inch steam-cylinder and a 3-inch to a 12-inch pump; these sizes will deliver respectively 1,300 and 38,400 gallons per hour to moderate heights. The stroke of the smallest size is 6 inches, and of the largest 28 inches. The number of revolutions per minute for the smaller sizes varies from 70 to 85 (105 feet

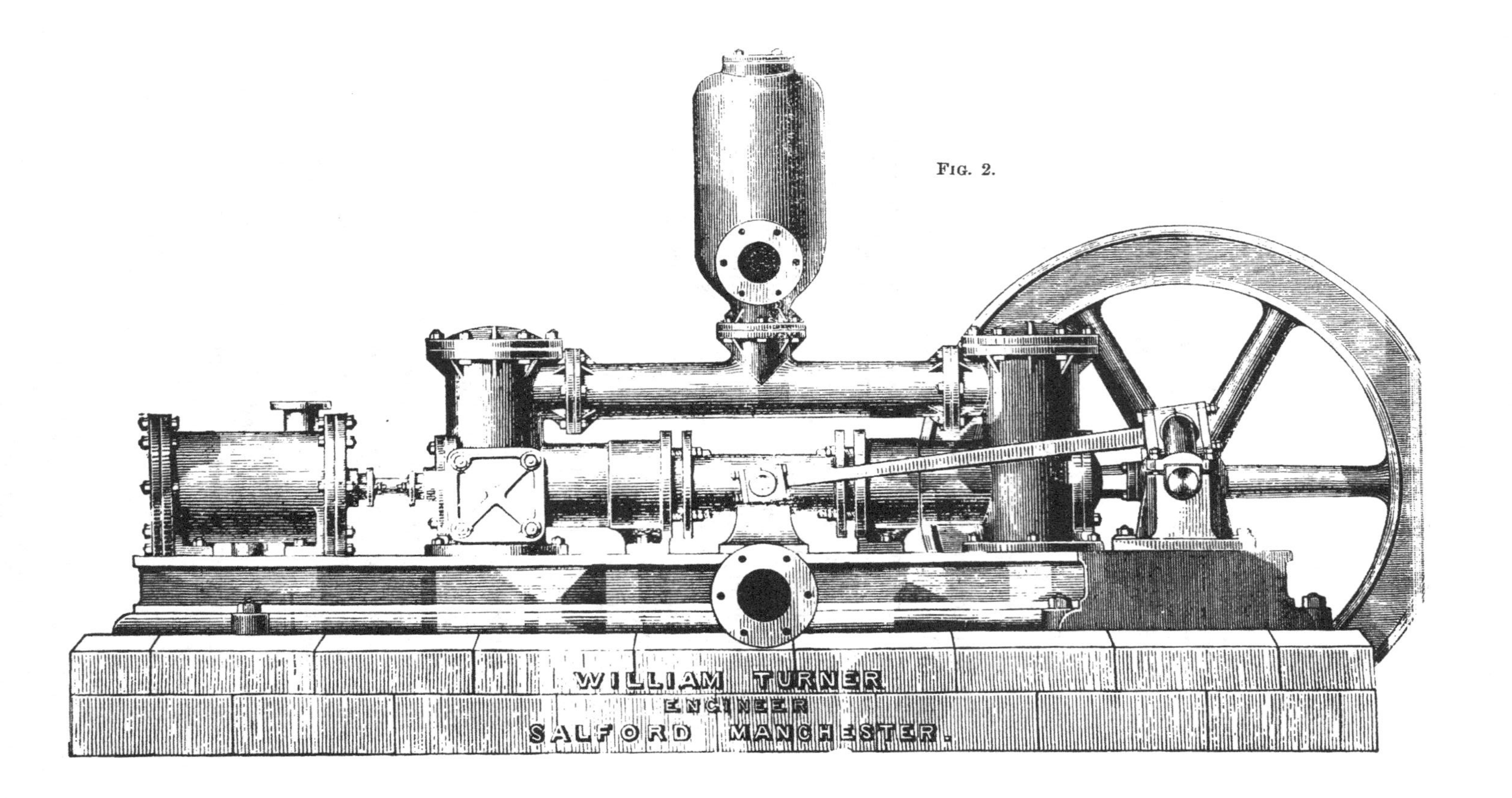

FIG. 2.

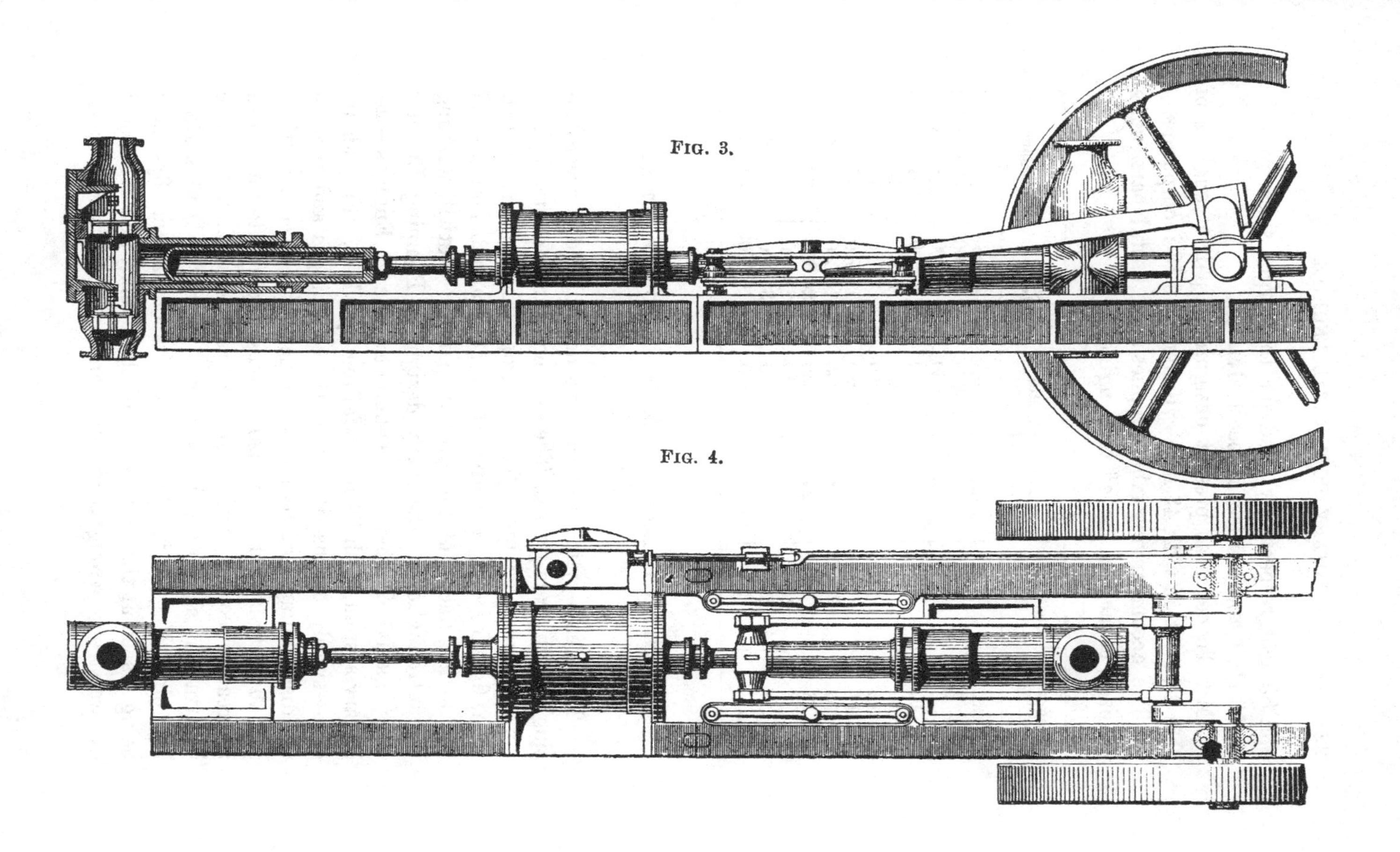

Fig. 3.

Fig. 4.

and 85 feet respectively), and for the larger sizes from 34 to 50 (160 and 125 feet respectively).

Mr. Turner is at present designing a pair of engines of this description with 43-inch cylinders and 13-inch pumps.

The accompanying table supplies some useful particulars of sizes, capacities, &c. of this engine.

Diameter of cylinder	Diameter of pump	Stroke	Height	Delivery per hour	Space required
in.	in.	in.	ft.	gallons	ft.
6	3	6	230	1,300	6 × 2½
8	4	8	230	2,700	7½ × 2½
10	5	10	230	4,500	9½ × 3
10	5	15	230	5,200	13 × 3
12	6	18	230	7,800	15 × 4
14	7	21	230	11,500	16 × 4
16	8	24	230	16,000	20 × 4½
18	9	24	230	20,300	20 × 5
20	10	28	230	27,000	22 × 6
30	12	28	350	38,400	22 × 6
30	8	28	800	16,000	22 × 6

Turner's Rotary, Horizontal, Double-acting Ram-Pump.

The steam-cylinder of this engine is placed between two single-acting ram-pumps. The fly-wheels are mounted behind one of these pumps. The accompanying elevation and plan, figs. 3 and 4, will give the reader a sufficient idea of the arrangement of the various parts. 'Notwithstanding the introduction of the many designs of Pumping Engines—as Special, Tappet, and Differential Lever Engines—the Rotary system, with its never-failing fly-wheel, still holds its own—simply because it has all the advantages and none of the disadvantages of its rivals. It is simple in construction, and the action of its every part, being exposed to view, can be readily comprehended by the rudest mechanic. The illustrations shown are W. Turner's Improved Double Ram Pumping Engine for heavy lifts, its crank and fly-wheel being placed at the back of one of the pumps, and connected by two strong wrought-iron side connecting-rods to a crosshead attached to one of the rams. The valve-gear is direct and simple, one valve alone being in this case necessary, a second or cut-off being added in the compound system, where a greater expansion is necessary. The pump rams may be made either of cast-iron or brass. The pump valve-boxes are attached directly to the ends of the pumps; they have therefore the shortest and most direct communication with the pump barrels—a very necessary consideration where the water is forced to a high level. The pumps, being rams, are found of great advantage where the water is muddy or gritty; the grinding action common in all piston-pumps is impossible, as the rams fit only at the stuffing-boxes.

'The engine and pumps are bolted to two strong girder beds, making the engine complete within itself; strong foundations are thus avoided, and the engine may be placed on a light concrete foundation, or simply bolted to wooden beams.

'Mr. Turner has made many of these pumps for various lifts in mines ranging to 900 feet and has every confidence in recommending them for their durability and certainty of action.' (Extract from maker's catalogue.)

The following are particulars of the sizes, capacities, &c. of this type of engine.

Diameter of cylinder	Diameter of pump	Stroke	Height	Delivery per hour	Space required
in.	in.	in.	ft.	gallons	ft.
21	7½	24	440	10,800	25 × 6
30	9	36	630	16,200	32 × 7
42	10	48	1,000	21,000	42 × 7½

Turner's Rotary, Horizontal, Double-acting Piston-Pumps.

Mr. Turner is the maker of four varieties of piston pumps; figs. 5, 6, 7 and 8 are illustrations of three of them In the first variety, fig. 5, the engine has a piston working in the pump-cylinder with a rod passing through the latter. To that part of the rod which passes beyond the outside pump cover is fixed a crosshead to which is attached a pair of connecting-rods, connecting the fly-wheels to the pump. The fly-wheels are mounted between the engine and the pump.

More than two thousand of these pumps have been made for mines, water-works, irrigation, &c. The largest sizes made are as follows—

Dia. of cylin.	Dia. of pump	Stroke	Height	Delivery per hour	Space required	
in.	in.	in.	ft.	gallons	ft.	
18	$4\frac{1}{2}$	12	900	4800	10 × 8	A pair of engines
14	4	12	700	1800	9 × 4	

In the next variety, illustrated by fig. 6, there is a pair of fly-wheels mounted at one end of the machine behind the pump, whilst the other end of the bed-plate supports the steam-cylinder. The crosshead connecting the fly-wheels to the pump is fixed between the steam and pump-cylinders.

This engine is adapted to deep mine drainage.

The following are the sizes made, their capacities, &c.

Dia. of cylin.	Dia. of pump	Stroke	Height	Delivery per hour	Space required	
in.	in.	in.	ft.	gallons	ft.	
14	7	24	230	19,500	17 × 9	A pair of engines
16	9	32	180	17,000	22 × 5	
20	7	24	460	9,750	17 × $5\frac{1}{2}$	
26	8	24	600	12,000	18 × 5	
30	$11\frac{1}{2}$	24	400	25,500	20 × 6	
31	11	24	450	40,000	20 × 16	A pair of engines
34	13	24	290	63,000	20 × 16	,,

In the other type of Turner's piston engines, the fly-wheels are mounted at one end of the machine, behind the steam-cylinder, and the pump is fixed at the other end. On one end of the steam piston-rod is fixed a crosshead, to which is attached the connecting-rod of the fly-

FIG. 5

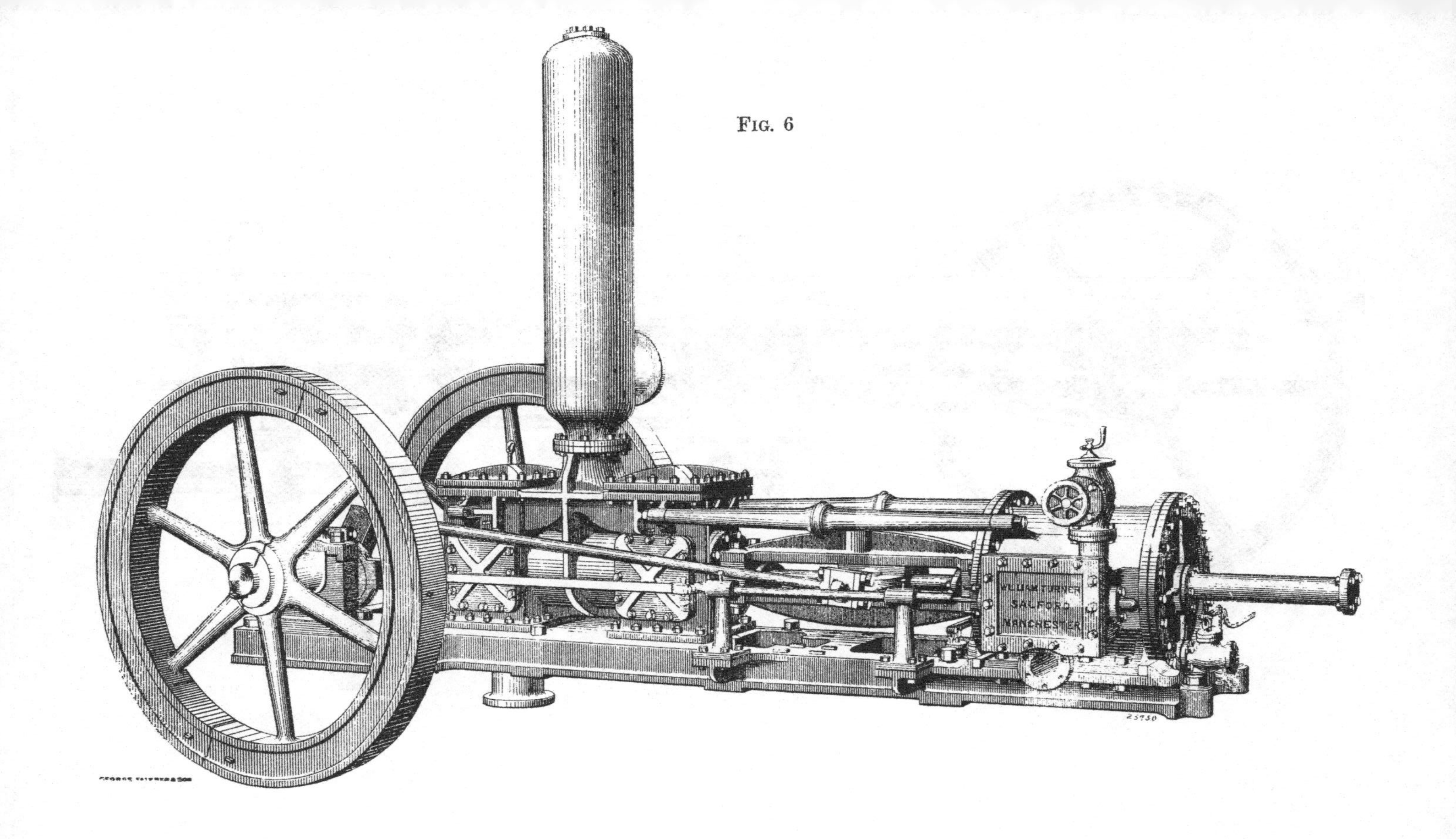

Fig. 6

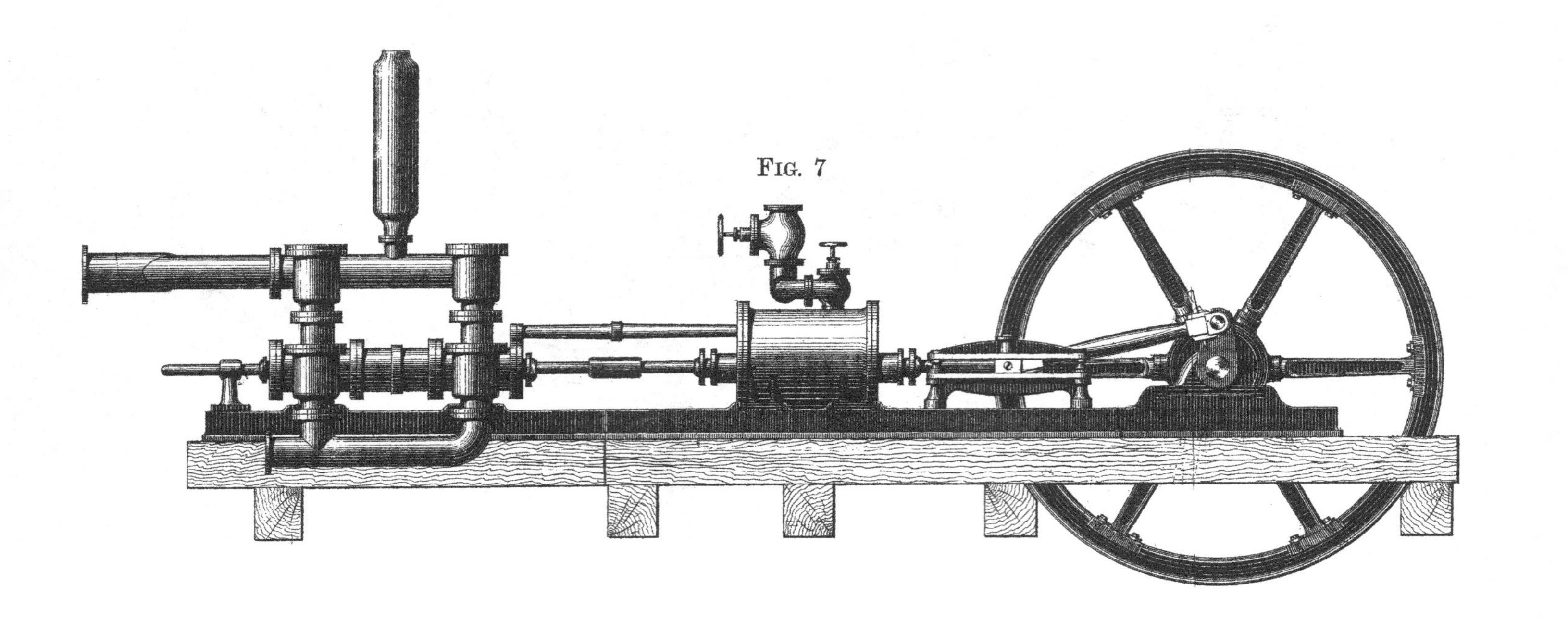

Fig. 7

FIG. 8

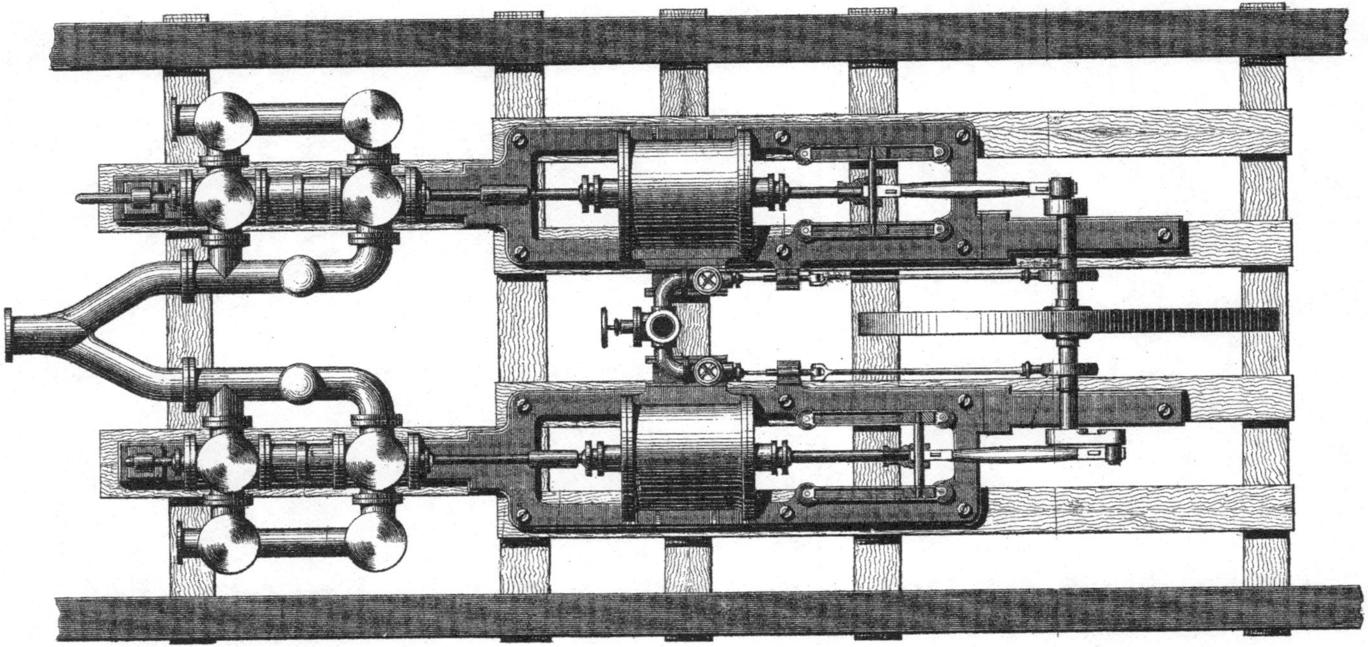

G. FALKNER & SON. MANCH^{R}

wheel; the opposite end is attached to the pump-rod, which passes through both pump-covers, the outside end working in a bushing guide, the object of which is to prevent the pump-piston bearing too heavily upon the bottom of the cylinder; this arrangement promotes its durability.

2. TWO STEAM-CYLINDER, EXPANSIVE, NON-CONDENSING ENGINES, WORKING DOUBLE-ACTING PUMPS BY CONNECTING-ROD AND CRANK, REGULATED BY A FLY-WHEEL.

TURNER'S ROTARY, HORIZONTAL, DOUBLE-CYLINDER PISTON-PUMP (QUADRUPLE-ACTING WITH TWO PISTONS).

It will be observed from figs. 7 and 8 that this engine consists of two steam-cylinders, each working a double-acting pump of the type last described. There is only one fly-wheel, to the ends of the shaft of which are fixed the cranks, set at right angles to each other, so that one engine shall be at its maximum point of power when the other is at its minimum, thereby securing an equable distribution of force throughout each stroke. There is an air-vessel on each pump, and the two delivery-branches discharge into the same delivery-pipe. Two pairs of these engines are at work in Germany for the Royal Mine Inspection; one pair at Zabreze, and the other pair at Königshütte. They have respectively 45-inch and 47-inch steam-cylinders, and 9-inch and 15-inch pumps, each having a 36-inch stroke. They are working against a head of about 900 feet. Mr. Turner has also made an engine of this type for Messrs. Stanier and Co., Silverdale, with a 43-inch steam-cylinder, 10¾-inch pump, and 36-inch stroke, and an engine for Mr. Waddle, Llanelly, with a 31-inch steam-cylinder, 12-inch pump, and 28-inch stroke.

The subjoined list contains interesting particulars of the sizes, &c. of this engine.

Dia. of cylin.	Dia. of pump	Stroke	Height	Delivery per hour	Space required	
in.	in.	in.	ft.	gallons	ft.	
21	13	28	150	33,780	20 × 7	
14	7	12	230	21,600	12 × 8	A pair of engines
31	12	28	380	57,600	20 × 12	,,
34	13	24	510	63,000	20 × 16	,,
43	10¾	36	910	60,000	34 × 16	,,
47	15	36	560	99,000	40 × 16	,,

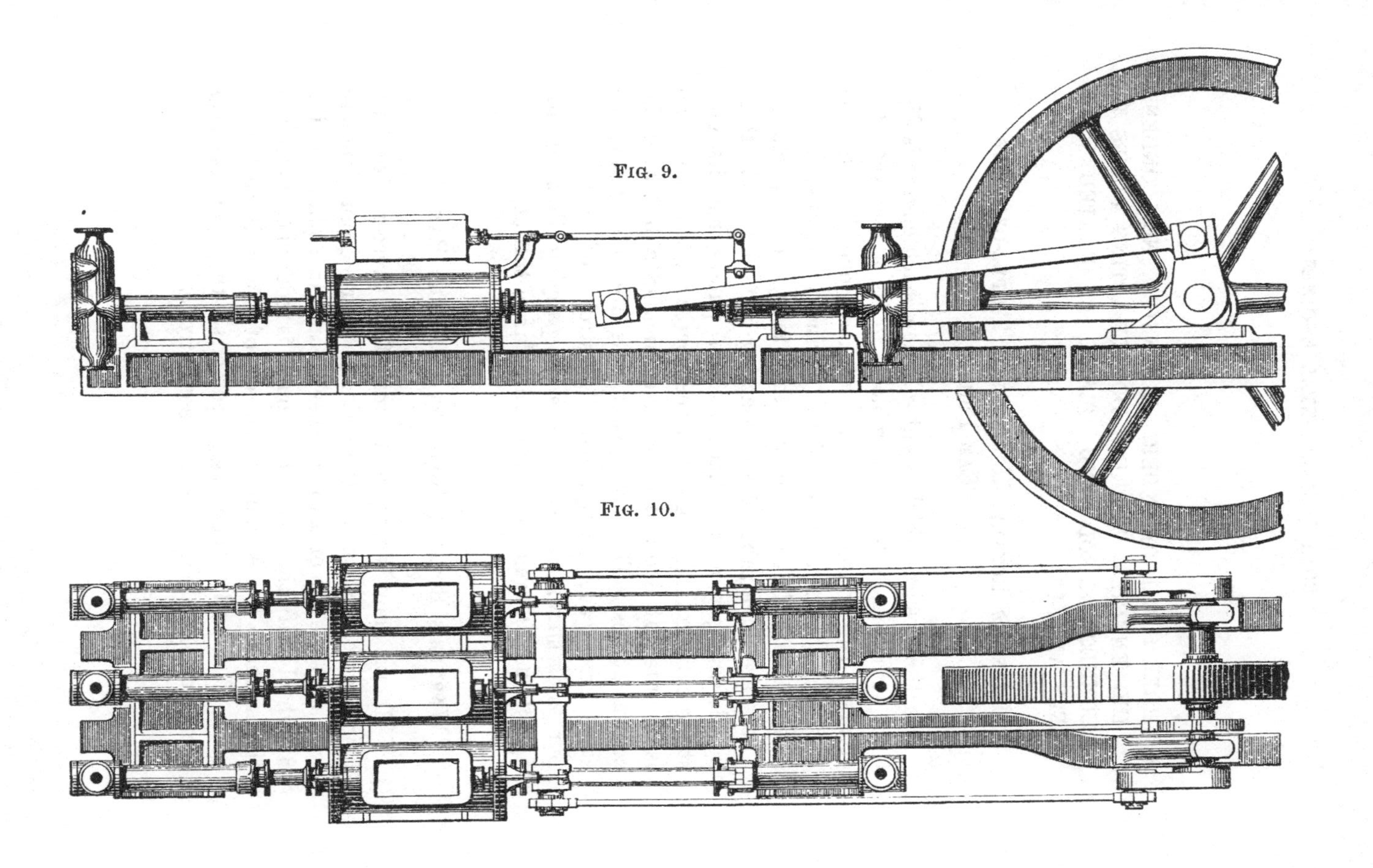

Fig. 9.

Fig. 10.

3. THREE STEAM-CYLINDER, EXPANSIVE, NON-CONDENSING ENGINES, WORKING DOUBLE-ACTING PUMPS BY CONNECTING-ROD AND CRANK, REGULATED BY A FLY-WHEEL.

Turner's Rotary, Horizontal, Triple-cylinder Ram-Pump.

In this design, which is entirely new, each engine is distinct within itself, but yet combined in its action with the others. The arrangement is an expensive one, but the fact that an accident to either of the engines or pumps does not interfere with the action of the others is of some importance. There are three pairs of single-acting ram-pumps (or three double-acting pumps). Each pair of pumps has its own steam-cylinder. A single fly-wheel serves for the whole, and is mounted at one end of the machine, in a line with the central pumps. The three piston-rods are connected to the same crosshead, consequently the stroke of each set of pumps is simultaneous. The accompanying illustrations, figs. 9 and 10, are an elevation and plan of this engine.

Warner's Rotary, Horizontal Triple Ram-Pump.

Messrs. John Warner and Sons, of the Crescent Foundry, Cripplegate, London, are the makers of a triple-ram pump, in which, by the use of a tooth-wheel geared in a pinion on the fly-wheel shaft, the speed of the pump is reduced below that of the engine, a corresponding increase of power in the pump being the result. The engine comprises three single-acting ram pumps, making a simultaneous stroke.

B. COMPOUND STEAM-PUMPS.

The compound principle of steam-cylinders is one which is rarely resorted to in engines of the rotary class. A reason for this is not far to seek. By using a heavy fly-wheel a great amount of expansion can be obtained with a single cylinder (by means of a cut-off steam-valve, working on the back of the slide-valve or otherwise, and actuated by a separate eccentric on the crank-shaft). If it suit the miner's purpose better to employ a rotary than a non-rotary engine, it will give, for small and medium size engines, sufficient expansion to prevent an extravagant consumption of fuel, and the extra expansion to be obtained by employing a second cylinder would not compensate for the increased size of the engine, the additional cost and the extra space required, and besides, as a portable engine it becomes a much less useful machine. With the non-rotary engines the case is quite different. These will give only a very low ratio of expansion under the most favourable circumstances, and then an initial pressure of steam largely in excess of the load and a great piston speed must be adopted. The compound principle in this case is an absolute necessity for large engines if the least regard be paid to economy. We have only observed one maker advertise compound rotary engines, which of itself is a good indication of the opinion of practical men on this point.

COMPOUND STEAM-CYLINDER, EXPANSIVE, CONDENSING ENGINES, WORKING DOUBLE-ACTING PUMP BY CONNECTING-ROD AND CRANK, REGULATED BY FLY-WHEELS.

Turner's Compound, Rotary, Horizontal, Double-acting Engine.

Mr. Turner is also the maker of an engine of this class, a plan of which is given at fig. 11. It will be seen that the low-pressure cylinder is fixed behind the high-pressure cylinder. Two connecting-rods on the crank-shaft are attached to a crosshead, to which are also attached the piston and pump-rods. On the crank-shaft are also two eccentrics, the outer one actuating the steam cut-off valve and the inner the slide-valves of both cylinders. The illustration shows no condensing apparatus, but the engine may be arranged with an air-pump and condenser, driven by an extension of the piston-rod at the back of the steam-cylinder, or by a separate condensing engine. They may be arranged as piston or double-ram pumps. The principle of its action is precisely similar to simple pumps of the same class. Engines of this kind are little used.

Fig. 11

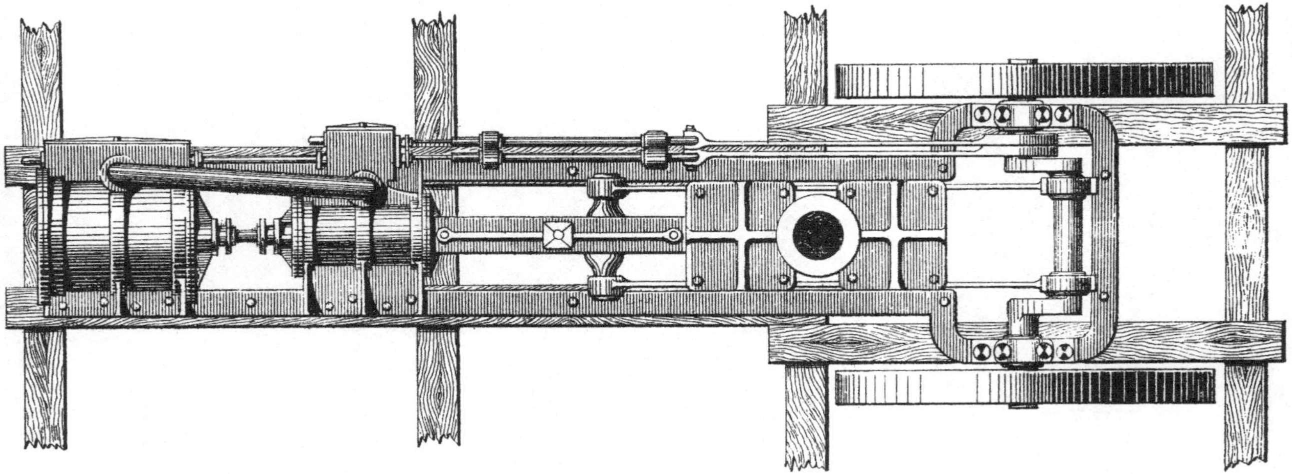

B. *NON-ROTARY HORIZONTAL ENGINES.*

A. SIMPLE STEAM-PUMPS.

WITH a few exceptions the pumps of this class strikingly resemble each other. Between some pumps the difference is very little indeed—a slight modification at some point, chiefly in the valve arrangement, sufficient to avoid an infringement of any existing patents. In other instances the difference is more marked, involving, in fact, quite distinct principles of action in the valve-gear.

It will be sufficient at present to indicate by a brief outline the general nature of the valve-gear, the mode of operating it, and the general arrangement of the pumps.

The pump has two cylinders, usually on the same bed-plate—a steam and a water-cylinder, which are in a straight line and are connected by a distance-piece, the end flanges of which form the covers for both cylinders. The piston rod attached to the steam-piston passes through both covers and is also attached to the pump-piston, thus giving the two pistons an identical movement.

The steam-cylinder is surmounted by a cylindrical steam-chest containing a starting piston-valve, or piston-valves, for shifting the main slide-valve. This piston-valve is actuated either direct by steam from the main-cylinder through shooting-ports at or near the ends of the latter, or by tappets (levers) moved by the main-piston, with or without the interposition of an auxiliary-valve, or by reversing-valves in the main-cylinder ends, moved by the piston, which admit steam from the main-cylinder to either end of the starting-valve.

In one type of the 'Universal' pump (the long-piston short-stroke) the separate valve-chest is dispensed with. The main-piston, which is longer than the stroke, is hollow, and contains a cylindrical valve-chamber in which a long piston, constituting the slide-valve, reciprocates. The main-piston is open at the top and sides between the ends, which are steam-tight to the cylinder, and the latter performs the functions of a valve in opening and closing the ports and steam-passages.

In Colebrook's pump there is no starting-valve, and it is, as far as we know, the only pump without small valves or tappets, and that has only one valve, which is of the piston type and in equilibrium, being actuated by the exhaust steam.

In Mr. Warrington's 'Standard' pump a circular piece or plug in the slide-valve directs the live steam coming from the main cylinder to either end of a cylindrical slide-valve.

The main-valve is usually of the D type, and is generally shifted by the starting-valve mechanically. In a few instances it forms a part of the starting-valve, partaking, thereby, of its motion. In the 'Excelsior' and 'Caledonia' pumps the main and starting-valves consist of a cylindrical spindle, two pistons near the centre forming the main-valve, and a double-headed piston at each end forming the two starting-valves.

In a few cases the main-valve is moved by the action of steam admitted to it by the movement of the starting-valve. The main-valve may thus be moved *mechanically* by the starting-valve or by the *direct impact of steam.*

In some pumps there is an auxiliary-valve underneath the main-valve (in addition to the starting-valve) operated by a tappet or tappets struck by the main-piston at the end of its stroke, by which steam is admitted to the starting-valve which gives motion to the main-valve.

In the 'Blake' pump a piston or plunger-valve, operated by steam from the main-cylinder, has the main-valve fixed between its two piston ends, and so moves it. An auxiliary-valve, contained in a chamber in the bottom of the steam-chest, and below the main-valve, and operated by tappets struck by the main-piston, supplies steam to the starting-valve, which carries the main-valve, and the latter gives the main-piston full steam.

The auxiliary-valves may be external to the cylinder as in

the 'Imperial' pump. They are operated by steam transmitted to them by the movement of the main-piston, and the steam passes to either end of a plunger starting-valve, connected to which is an ordinary flat valve which gives full steam to the main-piston.

The pumping part of the machine comprises, except in the case of ram-pumps, a large chamber, the bottom portion of which consists of a cylinder. The upper portion of the chamber contains two delivery-valves, one at each end, and surmounting the chamber, and midway between the valves, is an air-vessel into which the delivery-valves discharge. The upper part of the chamber is a separate casting, and is removable for examination of the valves, &c. A pair of suction-valves in the bottom part of the chamber admit the water to each end of the pump-cylinder alternately, which, it must be added, does not extend the whole length of the chamber, in order that the piston may communicate with each set of valves. The *modus operandi* is as follows:—A stroke of the piston to, let us say, the right-hand end of the cylinder will force the water which fills the space between the suction-valve, the piston, and the delivery-valve, through the latter into the air-vessel, the pressure of the fluid keeping the suction-valve closed. Meanwhile, the opposite end of the chamber having been emptied more or less by the previous stroke, the suction-valve immediately admits water into the space between it, the retreating piston, and the delivery-valve. At the completion of the stroke of the piston to the right, the right-hand end of the chamber is emptied of a great or the greater portion of the water it contained, and the left-hand end of the chamber and the cylinder as far as the piston are full. The return stroke of the piston forces a large portion of this body of water through the delivery-valve at that end (the left-hand) of the chamber towards which the piston is moving, the right-hand end of the chamber simultaneously filling. The displacement of water by each stroke of the piston is, of course, a volume the length of the stroke and the diameter of the cylinder, and these being known, the quantity of water pumped at each stroke can be easily calculated, slip being allowed for. From what has been said it will be gathered that these pumps are double-acting, that is, they force water at each stroke. The sections given further on will enable the reader more clearly to comprehend this fact, and will explain better than can be done by words the nature of not only the water part of the class of pumps under consideration, but also give him a better general idea of the steam part, and the pump as a whole, although, as we have already pointed out, each pump has its own particular method of arranging the valve-gear and other parts. The principle which secures the double-action is the same in nearly all these pumps.

Instead of a piston for the water-cylinder, a plunger or ram is sometimes used, but in general the piston is used, although for pumping gritty water the plunger is much superior. It works through stuffing-boxes, which, while permitting the plunger to move water-tight, prevent the wear and tear and derangement that attend the use of the ordinary piston in impure and gritty water by the friction between the piston and the cylinder.

'Piston-pumps are to be condemned (save in exceptional cases), especially in mines. A perfect pump for a permanent engine should have no internal packing or wearing surface which may be affected by grit or sand in the water, and a pump to fulfil these conditions must have a plunger or plungers provided with external stuffing-boxes. A piston-pump may be allowing a large quantity of water to pass the piston every stroke without its being detected; but with a plunger-pump the slightest leakage is at once seen. The stuffing-boxes of pumps should be packed with good hemp gasget steeped in melted tallow. There is no better packing.'

In concluding this outline we might add that these pumps are usually very compact, the two cylinders generally standing upon the same bed-plate. The total length and width of the medium sizes are within 13 feet by 5 feet, and of the small sizes 5 feet by 1½ foot. The largest sizes are imposing engines, extending to 18 feet by 7½ feet. A 'Universal' pump, 7-inch steam-cylinder and 5-inch pump-cylinder, occupies but 4 feet in length, and in height only 13 inches for two-thirds of this length, and 2 feet 8 inches for the remainder.

Having given an outline of the nature, construction, and arrangement of these pumps, we will proceed to lay before the reader some particulars of their sizes, weights, capacities, capabilities, cost, mode and circumstances of use, rates of working, and duty. Any account of the machines which omitted to consider such points would be of little use to practical men, who require to know the suitability or otherwise of certain machinery for their purpose by a consideration of such particulars relating to them as we have indicated and which we now begin to detail.

Sizes, Weights, and Speed.

Sizes and Weights.—These pumps vary in size from a steam-cylinder of 1½ inch to 40 inches, and a water-cylinder of 1 inch to 22 inches, and a stroke of from 6 inches to 6 feet. The very small sizes are used for tanneries, chemical-works, &c. The smallest pumps used in mines and collieries are from about 6 to 8 inches diameter steam-cylinder, and about 3 inches water-cylinder; even the smallest sizes have been in a few instances used in mines and collieries for feeding boilers or some other surface work of a very light character. A pump with less than a 9-inch steam and a 6-inch pump-cylinder would be of little service for underground pumping operations, except for emptying dips where the influx of water is small. Such a pump as the foregoing, with a piston speed of 100 feet per minute, would deliver over 7,000 gallons per hour.

The proportions of the sizes of the steam and water-cylinders vary according to the height the water has to be raised and the pressure at which the steam is used. Any combinations can be made between the steam and water-cylinders. For lifting against a head of say 200 feet, a fair proportion for a small pump would be 12 s.-c. × 8 w.-c.; for about 500 feet say 12 × 5, whilst still higher lifts would need say 12 × 3. The proportions of large pumps would be in the same ratio; thus for the above heights, say 26 × 13, 26 × 9, and 26 × 7 respectively. For the highest lift, from 900 to 1,000 feet, the ratio between the diameters of the steam and water-cylinders would be as 4 to 1.

The two following tables of the sizes and weights of some of the 'Universal' and 'Excelsior' pumps respectively, will afford an idea of the dimensions of pumps of this class, of which we think they may be taken as fair examples:

Diameter of steam-cylin.	Diameter of water-cylin.	Length	Width	Weight
in.	in.	ft. in.	ft in.	tons cwts. qrs. lbs.
4	2	3 0	0 10	0 1 2 20
6	3	3 8	1 5	0 3 2 0
7	6	4 4	1 5	0 6 2 10
9	6	4 10	1 5	0 8 1 14
12	4	8 6	3 6	1 5 0 0
12	12	9 0	3 6	1 18 0 0
15	6	10 10	4 2	2 5 0 0
18	4	11 3	4 2	2 8 0 0
18	12	8 10	3 1	2 15 0 0
18	15	9 6	4 0	3 10 0 0
21	5	12 6	4 6	4 0 0 0
25	4	14 0	4 0	5 0 0 0
30	8	14 0	4 6	8 0 0 0
40	13	18 0	7 6	13 0 0 0

Diameter of steam-cylin.	Diameter of water-cylin.	Length	Width	Weight
in.	in.	ft. in.	ft. in.	tons cwts. qrs. lbs.
4	2	3 6	0 11	0 2 1 0
6	3	3 6	1 0	0 4 0 0
7½	6	5 10	1 8	0 11 2 0
9	6	5 10	1 8	0 14 0 0
12	6	5 10	1 8	0 18 0 0
12	12	9 6	5 4	2 15 0 0
16	6	10 0	3 0	2 10 0 0
16	12	10 0	5 4	3 15 0 0

Speed.—The piston may be driven to a speed of more than 200 feet per minute, or it may be worked to less than a foot. Very small short-stroke pumps may be run to 350 strokes (175 double-strokes) per minute. The usual speed at which these pumps are worked is about 120 feet per minute for all sizes. Some pumps, if worked very slowly, are liable to stop, owing to the valves getting centred. It is a prime test of the efficiency of direct-acting pumps to endeavour to work them so slowly that the piston crawls along, as it were. The maximum speed of a 4-inch × 2½-inch pump, 6-inch stroke, is 350 strokes per minute = 175 feet; a 6-inch × 3½-inch pump, with 8-inch stroke, 300 per minute = 200 feet; a 12-inch × 7-inch, 12-inch stroke, 200 per minute = 200 feet. To prevent a misunderstanding it will be necessary to explain that we have taken the stroke of the engines to be the travel of the piston from one end of the cylinder to the other only. It matters little, when speaking of double-acting engines, whether we regard the stroke as the return of the piston to the position it left, or only its movement from one end of the cylinder to the other. In the accompanying table, compiled from Messrs. John H. Wilson and Co.'s lists, the latter view is adopted. The first eight are piston-pumps; the others are ram-pumps.

Table of Maximum Speed of the 'Selden' Pump.

Diameter of steam-cylinder	Diameter of water-cylinder	Length of stroke	Maximum number of strokes per minute	Maximum speed
in.	in.	in.		ft.
3	1	5	350	146
4	2½	6	350	175
4	2½	6	300	150
6	3½	8	275	183⅓
6	3½	8	300	200
8	5	10	250	208⅓
10	6	12	200	200
12	7	12	200	200
8	5	10	200	166⅔
8	6	18	150	225
16	8	30	75	187½
18	10	36	60	180
24	12	48	50	200
30	15	60	40	200
30	15	72	30	180
36	18	72	30	180

We observe from this table that the greatest number of strokes varies from 30 to 225 per minute, according to their length, and the piston speed from 150 to 225 feet. This may be considered to fairly represent the greatest capacity of pumps of this class. The usual working speed would probably not exceed two-thirds of that given.

As the friction of water through pipes increases in the proportion of the square of the velocity, it is obvious that the pump cannot be driven at a high speed without a serious loss of power. The speed of water through pipes ought not to exceed 200 or 250 feet per minute, and the pipes should be of large diameter. Large engines may be run faster with advantage than small engines, as the loss in friction is much less in proportion to the quantity of water pumped. A speed of 100 feet a minute is quite sufficient for small steam-pumps if an excessive resistance in the rising-main is to be avoided.

The rising-main for these pumps consists of cast-iron flange-pipes of about 9 feet long for large pipes, and 6 to 7 feet for small pipes. The sizes vary according to the size of the pump. The suction-pipe in a few instances is slightly larger than the discharge-pipe. The following table gives the sizes of suction and delivery-pipes for medium and high lifts for different size pumps. In the 'Special' and the 'Imperial' pumps the suction and delivery-pipes are each half the area of the water-cylinder.

Diameter of steam-cylinder	Diameter of water-cylinder	Diameter of suction-pipe	Diameter of discharge-pipe
in.	in.	in.	in.
5	3	2	1½
6	4	2½	2½
8	5	3	3
10	6	4	4
12	7	5	5
14	8	6	5
16	9	7	6
18	5	3½	3½
18	12	9	8
20	14	12	10
21	5	3½	3½
12	6	4	4
12	8	6	6
16	8	6	6
18	10	8	8
24	12	10	10
27	12	9	9
30	9	7	7
30	15	12	12
32	10	8	8
36	18	15	15

Capabilities.

These pumps can be made of sufficient size to meet the greatest requirements. The largest sizes will discharge a very large quantity of water a great height. A machine of 32-inch steam-cylinder and 14-inch water-cylinder will raise 40,000 gallons per hour a height of 313 feet, with a pressure of 40 lbs. per square inch of steam or compressed air; with a suitable increase in the size of the steam-cylinder this quantity of water could be raised to a much greater height, and if the pressure of the steam be also increased, the efficiency of the pump would be still further augmented. A pump of 30-inch steam-cylinder and 15-inch water-cylinder will deliver 60,000 gallons 100 feet, 57,000 gallons 120 feet, 54,000 gallons 140 feet, 51,000 gallons 160 feet, 48,000 gallons 180 feet, 45,000 gallons 200 feet high, with 25, 30, 35, 40, 45, and 50 lbs. of steam per square inch respectively.

Although the principle upon which these pumps are constructed can be carried to any extent—and makers advertise them to as large a size as a 40-inch steam-cylinder—we are of opinion that they can only be used with advantage for permanent work when the capacity of the machine is small, owing to the very great waste of steam which attends the use of large pumps. The waste in small pumps, although of course proportionately as great or greater, does not form so serious an objection to their use, and would be compensated by their great utility for special kinds of work.

When the drainage is of such a magnitude as to require an engine of a large capacity, the compound-cylinder pumps, which will form the subject of another article, and which have some claim to be considered economical, should certainly be employed in preference to the simple pumps.

In very deep pits two or more pumps could be used at different parts of the pit, one being placed at or near the water-level, another a certain height above it, and, if necessary, a third still higher, the pumps supplying each other in the same way as Cornish lifts.

A 'Special' pump supplied in 1871 to the Adelaide Colliery, Bishop Auckland, with a 26-inch steam-cylinder, 6½-inch water-cylinder, and 6-feet stroke, discharged 8,000 gallons per hour to a height of 1,040 feet in one direct lift.

A 'Special' supplied in 1873 to the West Yorkshire Iron and Coal Company, near Leeds, raised 16,000 gallons per hour 465 feet in one lift. At the Gosforth Colliery, Newcastle-on-Tyne, a pair of 32-inch engines of the same type, working 7-inch pumps, with 6-feet stroke, will raise 18,000 gallons per hour 1,068 feet, with a steam-pressure of 33 lbs. per square inch.

At the Foxhole Colliery of the Patent Anthracite Coke Company, Swansea, a 'Universal' pump of 21-inch steam-cylinder and 6-inch water-cylinder has been in use pumping 3,500 gallons per hour 400 feet vertically; steam-pressure 40 lbs.

Three 'Universal' pumps at the Gelli Colliery, Rhondda Valley, have done the following work:—1st pump, 352 yards down the slant, 12-inch steam-cylinder, 6-inch pump, raises 10,000 gallons per hour 117 feet high through 485 yards of pipe—steam on boiler at surface 40 lbs.; 2nd pump, 12-inch steam-cylinder, 6-inch water-cylinder, lifts 18 feet vertically through 120 yards of pipe—steam carried 494 yards; 3rd pump, 15-inch steam-cylinder, 7½-inch water-cylinder, raises 17,000 gallons per hour to a vertical height of 156 feet through 388 yards of piping—steam carried 406 yards. In all cases the exhaust steam was condensed.

At the Yate Collieries, near Chipping Sodbury, a 'Universal' has lifted 2,500 gallons per hour 480 feet, at 30 strokes per minute, with steam-pressure at 30 lbs.; boiler 340 yards from the pump; the size of the pump was 15 inches × 4 inches.

The value of these pumps in an emergency, such as the flooding of a mine, was strikingly demonstrated in the terrible inundation of the Tynewydd Colliery in April 1877, when pumps of the 'Special' type were the chief means used to overcome the vast body of water which had burst in upon the workings from an adjoining mine, and without which the rescue of the entombed men would have been impossible. Some account of the action of these pumps at the above-named colliery, visited by the author a short time after the accident, will be useful as showing their value in a disaster such as occurred there. The 6-inch bucket-lift in use prior to the accident for draining the dip workings was choked, and could not be set to work. A 'Universal' pump lent by the Havod Colliery was immediately taken down the pit, and fixed near the water-level. It had a steam-cylinder of 12 inches and a water-cylinder of 6 inches, with a 14-inch stroke. It was run to an average of 80 strokes per minute, the greatest speed being 120 strokes. This pump was thrown aside on the arrival of a larger and more efficient 'Special' pump from the Llwynypia Colliery, and which became the main instrument in clearing the dip. It had a steam-cylinder of 12 inches and a water-cylinder of 10 inches, with an 18-inch stroke. It was worked at an average speed of 40 strokes per minute, the greatest speed being from 45 to 50, and the lowest 35 strokes. At the bottom of the pit, which is 88 yards deep, was another 'Special' pump assisting a Cornish double-plunger-lift of 12 inches diameter and 3-feet stroke, worked by a large water-wheel, to raise the water pumped into the sump by the steam pumps above described. The 'Special' in the dip worked without intermission, save lowering it twice to keep it near the water (the 'Universal' being worked whilst these removals were being made), 12 days, pumping about 50 tons (11,000 gallons) per hour. The 'Universal' worked 3 days, pumping about 30 tons (6,700 gallons) per hour.

These pumps were supplied with steam by two locomotive boilers in a roadway on a level with the top of the dip and the bottom of the pit. The top of the dip is 720 yards from the pit, and extends to about 143 yards, at an average inclination of about 5½ inches to the yard. The highest point reached by the water was about 3½ yards vertically from the roadway. The large 'Special' pump in the pit, capable of lifting 13,000 gallons per hour, was supplied by steam from a locomotive boiler at the surface, which supplied the steam at 110 lbs. per square inch, and it was used at 102; 8 lbs. being lost in transit. The steam-pipe was 2½ inches in diameter and 130 yards long. This pump was worked about 12 days.

The small 'Special' steam-pump introduced at New Cook's Kitchen Mine a few years ago, where it is used as a boiler-feeder, is the only instance that we know of steam-pumps being employed in the mines of Cornwall.

The great cost of fuel prevents the use of these pumps in Cornwall for mining purposes, at least on a large scale. We believe they might be advantageously applied as temporary auxiliaries to the more economical Cornish engines, for draining distant points not easily accessible to the pump lifts, such, for instance, as winzes and other workings. It would be easy, we think, to work an air-compressor underground by deriving the power for compressing the air from the main-rods of the Cornish engine, and conveying the compressed air through the level to the steam-pump wherever the services of the latter may be required. They could also be used advantageously in sinking trial shafts. Many cumbersome and unwieldy, as well as costly, appliances are sometimes used for draining new shafts after they have become too deep to be drained by the tackle. When it is necessary to resort to steam-power we know of no better vehicle for its use than the steam-pump. Should the sinking of the shaft lead to important results, and issue in the development of a deep and extensive mine, the steam-pump could be replaced by the Cornish engine and pumps, and the former will be available for similar services elsewhere, or as an auxiliary to the Cornish engine.*

* The following remarks of Mr. W. Husband on the use of auxiliary engines underground, which the author only had the pleasure of seeing after these papers went to the press, are well worthy of quotation as the opinion of a gentleman eminently qualified to speak on the subject: 'In sinking wet shafts considerable difficulty is frequently experienced in keeping the pump clear of water so as to allow the men to work. In fact, where the water is quick and the lifts large, a considerable quantity must always accumulate in the bottom of the shaft, although the holes in the windbore be plugged low down, and the power required to draw the water through the contracted water-way thereby be injuriously increased. Under such circumstances the operation of sinking would be greatly facilitated,

Since the foregoing was written a small 'Special' pump of 5-inch steam and 2½-inch water-cylinders and 10-inch stroke has been introduced at Dolcoath, near Camborne, Cornwall, probably the deepest and richest tin mine in the world, to drain a winze from the 338 to the 352-fathom levels. It is worked with compressed air, of about 50 lbs. pressure per square inch, supplied by a powerful compressor at surface recently constructed by the Perran Foundry Company from the designs of Mr. Loam for working the boring machine.* The author also learns that a steam-pump was used underground at East Pool Mine a few years ago, and that one is in use at Levant Mine, in the submarine part.

These pumps are placed over or alongside a sump or lodgment where the water collects, and into which the suction-pipe is dropped. They may be worked by steam, which is usually the case, or compressed air, but as the consumption of coal necessary to produce a given quantity of compressed air is much greater than to produce the same quantity of steam at a like pressure, it is much cheaper to use the latter.† The steam may be taken from a boiler on the surface, or from one underground near the pump. The compressed air is taken from the surface. For sinking shafts or pumping out old mines the pump is carried upon a rough timber framing, slung in the shaft with a chain (as shown in the illustration

and the strain on the pump-work diminished, by the employment of a small auxiliary pump, to run about 20 or 30 strokes per minute, and lift the water into a cistern about three or four fathoms above the bottom of the shaft, the main pump to lift the water from this cistern. The auxiliary pump could be driven by compressed air, which might be obtained by a compressor worked off the main rod, or by other means. Should the shaft be sunk by boring machinery, compressed air would be available for the purpose ; a flexible suction-pipe might be employed for the auxiliary pump, which could be easily handled and removed when blasting' ('Proceedings of the Mining Institute of Cornwall,' vol. i., No. 5, p. 168).

* This pump is now (May 1880) being used for draining a winze below the 352-fathom level. The author is indebted to the manager, Captain Thomas, for these particulars.

† 'The attainable limit of the useful effect of compressed air is about 50 per cent. of the power exerted in compression' (Dr. Siemens). 'The

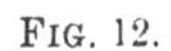

Fig. 12.

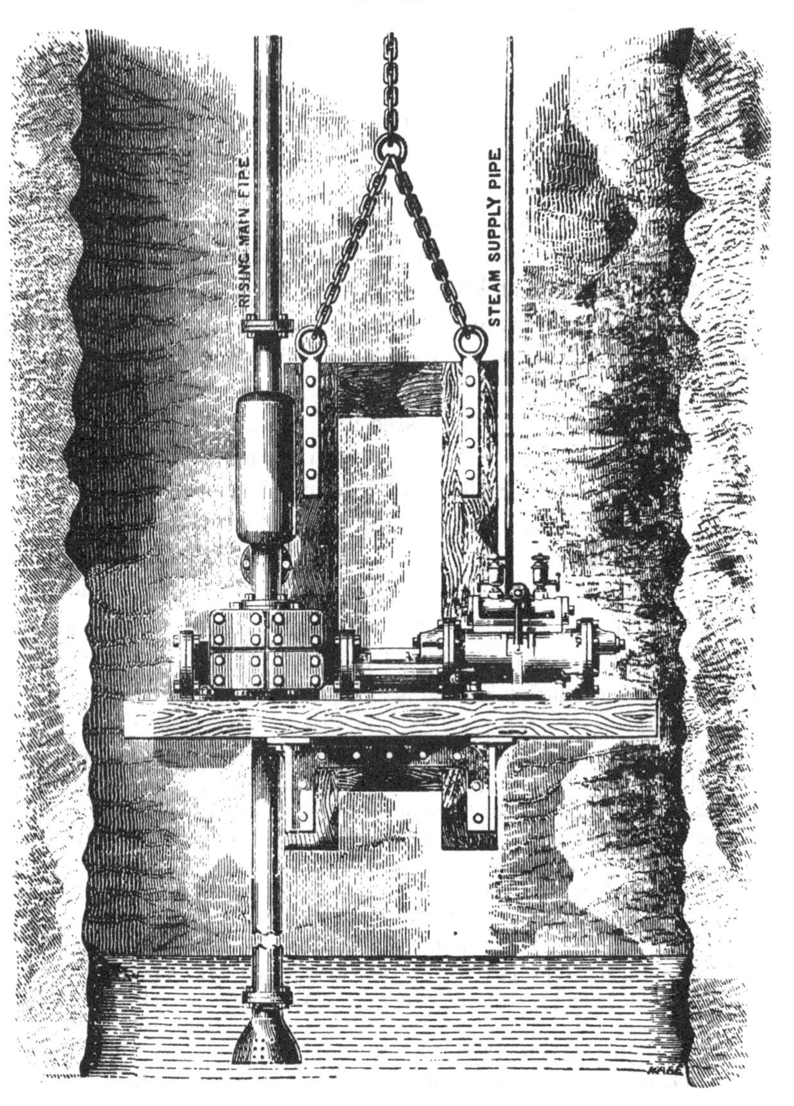

fig. 12, which represents a 'Special' pump), and lowered, by means of a crab or capstan, from the surface as the work of sinking or pumping proceeds. When slung, the pump and framing occupy but a small space at one side of the shaft, leaving the rest of the shaft clear for sinking and hauling operations. The lengths of rising-main-pipe are each secured to the sides of the shaft as the work proceeds, so that when the pump is lowered additional lengths of pipe can be applied easily.

The steam-supply-pipe, which is recommended to be of screwed wrought-iron, can be fixed and lengthened in the same way as the rising-main-pipes. We have hitherto spoken only of the merits of these pumps for mine drainage because their capabilities for this purpose only concern us and demand our consideration, but we might say that they have been successfully applied to pumping in breweries, tanneries, chemical-works, water-works, gas-works, sewerage-works, &c., and for feeding boilers.

Cost.

It will be useful to know the cost of these machines. The prices of different makers for the same size machines vary a little, but the following is a list of the general prices for the sizes given. An extra charge is made for a brass-lined, solid brass, or gun-metal water-cylinder and copper air-vessel. The table shows the extra cost for a condenser, &c.

loss of power seldom amounts to less than from 65 to 75 per cent.' (Professor Rankine). 'Compressed air has been used to work coal-cutting machines at Messrs. Baird's works at Gartsherrie. In this case 2½ cubic feet of steam, at 40 lbs. pressure, gave 1 cubic foot of air at 50 lbs. pressure. Compressed air has been used since 1864 in the shops of Messrs. Eastons and Anderson at Erith, where the consumption of coal necessary to produce a given quantity of compressed air was found to be about 69 per cent. more than to produce the same quantity of steam at a like pressure.' (*The Transmission of Motive Power to Distant Points*, by Mr. H. Robinson, M.I.C.E.)

Diameter of steam-cylinder	Diameter of water-cylinder	Stroke	Price	Price with condenser, blow-thro'-valve, and exhaust to condenser
in.	in.	in.	£	£
7	6	12	45	61
8	6	12	50	66
9	6	12	55	71
9	8	18	70	94
10	7	12	65	88
10	10	24	95	135
12	8	18	85	112
12	12	24	140	190
14	7	24	110	138
14	12	24	160	215
16	8	24	145	168
16	14	24	195	255
18	9	24	180	225
18	14	24	230	290
20	16	36	335	—
21	12	36	270	330
24	12	36	285	345
26	14	48	400	470

The following table of prices, &c., of Davey's 'Differential' steam-pump is given *in extenso* because it furnishes other particulars which are useful:

Dia. of steam cyl.	Dia. of water cyl.	Length of stroke	Gallons per hour (approx.)	Price	Price with condenser in suction pipe	Price with air pump condens	Dia. of suction and delivery pipes	Dia. of steam pipe	Dia. of exhaust pipe	Height to which water can be raised with 40 lbs. boiler pressure
in.	in.	in.	gallons	£	£	£	in.	in.	in.	ft.
10	5	15	5,496	71	88	99	3½	1½	2½	255
10	7	15	10,776	77	96	107	5	1½	2½	130
10	9	15	17,820	82	104	115	6½	1½	2½	78
12	6	20	7,920	93	115	126	4½	2	2¾	255
12	7	20	10,776	99	128	139	5	2	2¾	186
12	8	20	14,076	104	132	143	6	2	2¾	143
12	10	24	24,000	132	173	187	7½	2	2¾	91
14	7	24	11,760	132	154	170	5½	2½	3	255
14	8	24	15,360	143	165	181	6	2½	3	195
14	9	24	19,440	165	187	203	6½	2½	3	154
14	10	24	24,000	154	198	214	7½	2½	3	125
14	12	24	34,560	181	231	247	9	2½	3	86
16	8	24	15,360	165	187	209	6	3	3½	255
16	9	24	19,440	176	209	231	6½	3	3½	201
16	10	24	24,000	181	231	253	7½	3	3½	162
16	12	24	34,560	209	253	275	9	3	3½	113
16	14	24	47,040	242	275	297	10½	3	3½	83
18	9	28	19,440	209	253	280	7	3½	4	255
18	10	28	24,000	220	264	291	7½	3½	4	206
18	12	28	34,560	242	286	313	9	3½	4	143
18	14	28	47,040	264	308	335	10½	3½	4	105
18	16	28	61,440	297	341	368	12	3½	4	80

Duty.

The quality for which the Cornish engine is so renowned, and for which it has, perhaps, for deep and permanent pumping, never been surpassed—the high duty or work performed for a small consumption of fuel—is not possessed by any simple steam-pump, and, indeed, the makers do not profess that their machines are economical in this respect. These pumps, so far from being economical in working, are really very wasteful of steam, and the boiler capacity requisite for keeping them supplied with steam, taking into consideration the relative amounts of work performed, is exceedingly great compared with engines constructed on the most economical principles, such as the Cornish, Bull, Davey's

FIG. 13.

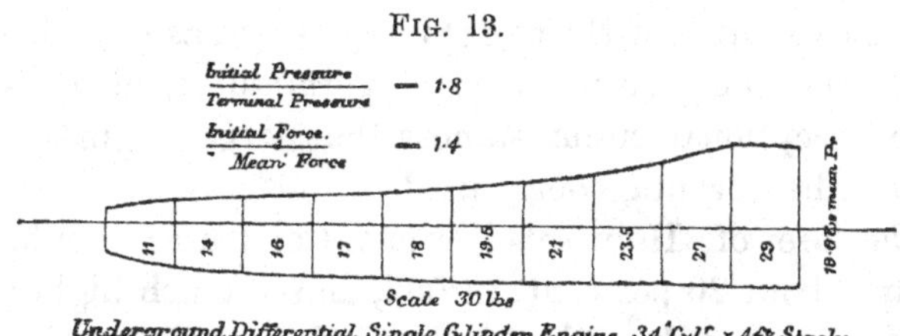

Underground Differential Single Cylinder Engine. 34″ Cyl.r × 4ft. Stroke

Compound 'Differential,' and other surface pumping engines. The indicator diagram, fig. 13, of a 'Differential' engine shows that the expansion in an engine of this class is very small, the ratios of initial and terminal pressures being only as 1 8 to 1. In a letter to the author the manager of the Tynewydd Colliery, where a 'Special' (steam-cylinder 12 inches, water-cylinder 10 inches, stroke 18 inches) is at work, says: 'We consume about 14 cwt. of bituminous coal (No. 3 Rhondda house coal) to raise a little over 300 tons of water to the height of 66 feet, but there is a certain quantity of heat lost, as the fires are taken out at night.'* This will give a duty of 3,168,000 lbs. raised one foot by the consumption of

* This pump is supplied with steam by two locomotive boilers placed underground at a distance of 120 yards from the pump, the steam being conveyed in a 2-inch pipe, and is exhausted to the atmosphere.

1 cwt. of coal; making an allowance for loss of heat as explained, and we have a duty of say 4,000,000, which is about one-twentieth of that performed in Cornwall by the best Cornish engines: or if we go back to the time when greater attention was paid to the engines and very much higher duties were performed, the difference will be far greater. The well-known Cornish engine erected by the Perran Foundry Co. at the United Mines in the year 1840, gave an average duty during the years 1841 and 1842, according to Mr. Farey, of 95,750,000 lbs., raised 1 foot by the combustion of 1 bushel of coal. Taking the bushel to be 94 lbs., this would be equal to a duty of 114,000,000 for 1 cwt. of coal, or nearly 33 times that performed by the steam-pump to which we have referred. It may be urged, and very justly, that the comparison of a small engine with a large one is not a fair criterion as regards the relative values of the two principles, but as regards actual results the comparison is quite permissible, as it is only under exceptional circumstances that large engines of this kind can be advantageously used.

The use of Holman's Patent Condenser would save perhaps about 20 per cent. in fuel, and a much higher duty than that recorded would be obtained, but the author does not think that under the most favourable circumstances the duty of pumps of this class would exceed 10,000,000 or 15,000,000 lbs. raised 1 foot by the expenditure of 1 cwt. of coal. It is impossible to make steam-pumps economical without a second cylinder, but then the lightness and simplicity for which they are noted will be gone.

In a paper * read before the Chesterfield and Derbyshire Institute of Engineers on October 2, 1875, by Mr. John B. Simpson, that gentleman assumes that the 'Special' pump will consume at the least from 10 to 12 lbs. of coal per effective horse-power per hour. The author is of opinion that engines of this class cannot be worked underground with less than 15 lbs. of coal per effective horse-power per hour and that under some circumstances it would exceed that quantity.

* 'Observations on the Comparison of the Cornish Engine and Holman's "Special" Pump.' By John B. Simpson, M.I.C.E., F.G.S.

Mr. Simpson's remarks are well worth considering, and we insert them presently with only a very slight curtailment.

In making choice of a steam-pump a long-stroke engine should be preferred to one with a short stroke, as it is more economical, because making fewer strokes for the same piston speed, it has to fill the clearances between the piston and cover a fewer number of times. For instance, a pump with a 2-feet stroke, running at a piston speed of 120 feet per minute, makes 30 strokes, and, consequently, has to fill the ports and clearances 30 times, whereas a 4-feet stroke pump, working at the same speed, would only have to fill them 15 times. 'It is undeniable that the majority of this class of steam-moved pumps have the fault of their use involving a comparatively great expenditure of steam; where the pump is small this may be hardly worthy of notice, but as it becomes of larger diameter and longer stroke, the economy in the consumption of fuel begins to be an important question, and the employment of such pumps has been thus limited to a certain class of cases. . . . This extravagance in the use of steam is a characteristic of direct-acting steam-pumps.*

'The great drawback to this class of pumps, hitherto, has been their inordinate consumption of steam. This is a fatal objection in the case of pumps of large size, for very soon the question of economy forces itself upon the notice of the user, and the consumption of fuel becomes a question which he is obliged to take into consideration. As there is no rule without exceptions, perhaps there are certain pumps to which these remarks do not so sweepingly apply, but that a perfect direct-acting pumping engine has been found, few, if any, will be prepared to admit.' †

The following are Mr. Simpson's remarks: 'I have read with pleasure Mr. Holman's description of the progress made in pumping machinery, and his statement respecting the Cornish engine. I can fully confirm what he has said of the duty and efficiency of the latter. I must say, however, that I felt considerably disappointed when I came to a de-

* *Engineering*, Sept. 7, 1877. † *Mining World*, Oct. 6, 1877.

scription of machinery which Mr. Holman intends to substitute for the Cornish engine, and which furnishes no details of the annual economy of his system of pumping. I expected that a comparison would have been instituted between the old and new systems. . . . Mr. Holman admits that his engine has not hitherto been an economical user of steam.

'Now, as an advocate of the Cornish engine for many years, and the means of its introduction into the North of England collieries, I am not disposed to admit that the engine in question has any claim to be compared with it. It would have added more to the practical utility of this paper had Mr. Holman given us the result of experiments in this respect; but to make up for this omission, I purpose to give an idea of the relative economy of the two engines.

'1st. *The Cornish Engine.*—Assume an engine exerting 500 horse-power of actual duty as mentioned by Mr. Holman. The quantity of coal it would consume annually, assuming the effective duty in water lifted at 4 lbs. per horse-power per hour, or about 50,000,000 duty, would be 7,821 tons, which at 5*s*. gives £1,955

'To this add the interest on cost of engine and pumps, which is estimated by the author at £7,000, say at 10 per cent. 700

Total per annum £2,655

'2nd. *Mr. Holman's Special Engine.*—This engine, working without any expansion or in any economical way as regards steam, will consume at the least 10 to 12 lbs. of coal per horse-power per hour of effective duty. I will assume 10 lbs. This would require 19,552 tons of coal per annum, which at 5*s*. will give . £4,888

'I do not know the cost of an engine of this description, but probably with the necessary pumps it would be £3,000 at least; then 10 per cent. on this amount gives * 300

Total per annum £5,188

* The extra number of boilers required, the greater cost of stoking, slate and other material, and repairs, must be added to this estimate.

'This estimate shows a difference in favour of the Cornish engine of £2,533 per annum, when coal is selling at 5*s.* per ton, and of course in times such as we have passed through I need not say how much more the difference would be. I maintain, therefore, as opposed to the author's views, that annual economy is more to be considered than first cost. I have not gone into the question of the labour employed, but I may say that the lesser number of boilers which would be required by the Cornish engine would cause a saving in this respect, and I must contradict the remark made by Mr. Holman, when he says that in his engine the labour is one-twentieth part of that of the Cornish engine.

'One more fact as to the Holman's engine. I understand it to be purely an underground engine, and in heavily watered sinkings it will still be requisite to have a surface engine to enable you to reach the point before the Special pump can be applied. In this case you have an increased capital, and at a heavy annual outlay, if you substituted the Special pump for the Cornish engine.

'There is no doubt that the Cornish engine in first cost is great, and that it has some drawbacks; but I have not yet been able to find any other engine for the draining of mines doing such economical duty, even taking into consideration the interest upon its first cost.

'I will freely admit that the Special pump for temporary purposes, and where economy is not an object, has been and is exceedingly useful, and especially where space is an object, and I think that colliery owners should give every encouragement to those engineers who are willing to give their attention and ability to enable the drainage of mines to be accomplished with greater facility and economy. I have no hesitation in saying that there is at the present moment four times as much coal consumed as there ought to be for this purpose. Any engine, however, that does not work expansively or on the compound principle, or is not designed with the object of economy of fuel, will not, in my opinion, be the engine of the future for the permanent drainage of our collieries.

'I cannot agree with the author of this paper that the Special pump which he has described is destined to supersede the Cornish engine. The arrangement of the valves and the little attention required I have no doubt are objects for admiration, and I fully believe all that Mr. Bigland has stated respecting them, and if Mr. Holman would endeavour to apply the principles of economy to the use of steam in the cylinders, and give us say 3 or 4 lbs. effective duty, then perhaps it might be the germ of that engine which he predicts will be the pumping engine of the year 1900.'

1. SINGLE STEAM-CYLINDER, NON-EXPANSIVE, NON-CONDENSING ENGINES, WORKING DOUBLE-ACTING PUMP.

THE 'SPECIAL' STEAM-PUMP.

This pump is the invention of Mr. Adam Scott Cameron, of the United States. The first engine was made in America late in the year 1865. It was introduced into this country in 1867 by Messrs. Tangye Brothers of Birmingham. It is the most widely known of its class, and is very extensively used, more than 12,000 being in use. As will be observed from fig. 14 (which shows how this engine may be made condensing) and fig. 15, this machine consists of a steam and a water-cylinder in a line with each other, and connected with a distance-piece the end flanges of which form the covers for both cylinders. The steam-cylinder, upon which is fixed the valve-chest, is made with a double set of steam-passages, the inner pair leading from the slide-valve face (E, in the section fig. 15) to the ends of the cylinder, and the outer pair extending from near the ends of the steam-chest to near the cylinder-covers. In each of these passages is fitted a reversing-valve (G G) which closes the opening to the cylinder. Except when moved by the piston these valves are kept against their seats by the pressure of steam on their backs. The outer ends of the valve-chambers have free communication with the steam-chest by small passages. The slide-valve covers the exhaust-port and one pair of steam-ports, and is so made that when it is removed to the right, steam is admitted into the right-hand port, and *vice versâ.*

On the back of the valve are a pair of lugs fitting between two collars, formed on a spindle connecting a pair of plungers (D D), which work in the cylindrical portions forming the ends of the valve-chest (C C), and into which the second pair of steam-ports (M M) open. The plungers (D D) are for the purpose of shifting the slide-valve, so as to permit ingress and egress of steam to and from either side of the

Fig. 14.

steam-piston alternately in the ordinary way, and are made to work comparatively free, so that sufficient steam will pass them to form a cushion at either end alternately.

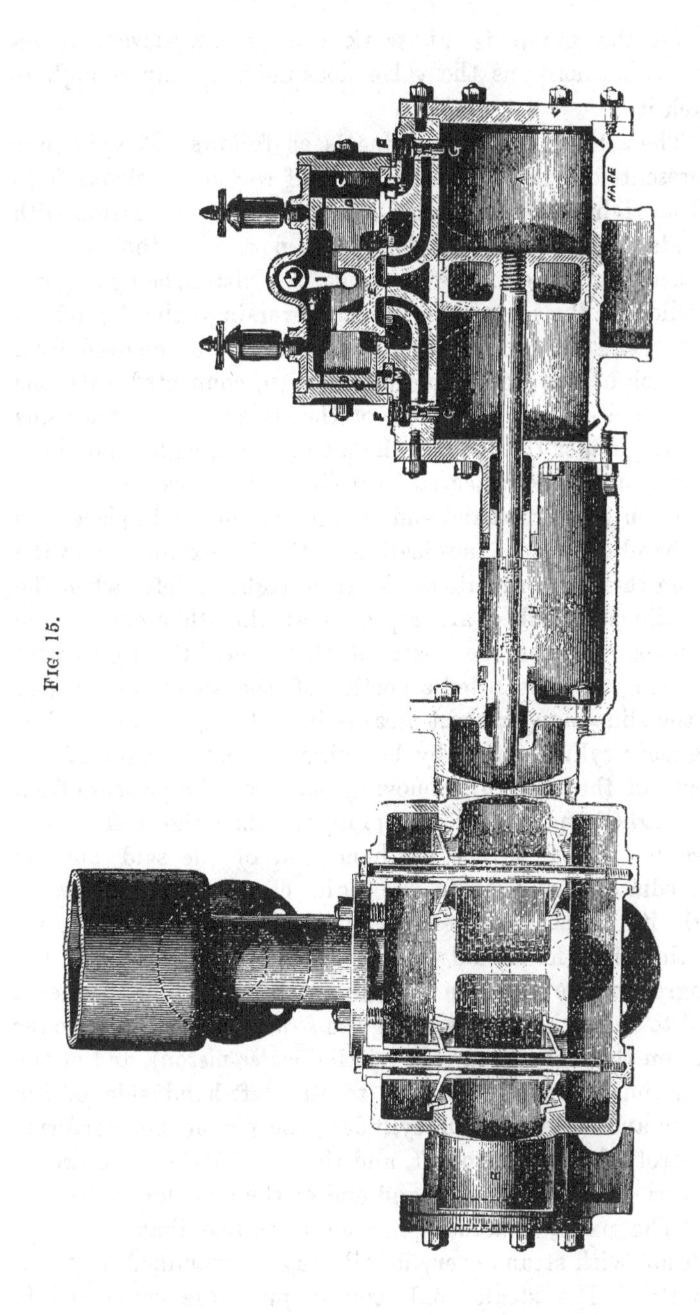

FIG. 15.

When the pump is at work the starting-lever (I) remains stationary, as the valve does not move far enough to reach it.

The action of the machine is as follows: The piston is represented in the diagram to be half way in its stroke from left to right. The left-hand port is in communication with the steam-inlet, and the right-hand port with the exhaust. On arriving at the end of its stroke, the piston, being slightly bevelled on the edges, will lift the reversing-valve (G) off its seat. This being the case, the pressure is removed from the back of the right-hand plunger (D) connected with the main-valve, and the pressure of the steam on the inner side of the plunger then forces the latter to the right, the slide-valve being of course carried with it. This movement admits steam on the right-hand end of the cylinder, and places the left-hand end in communication with the exhaust, and the piston then performs its stroke from right to left, when the operations described are repeated at the other end of the cylinder. It must be observed that when the right-hand reversing-valve is lifted a portion of the steam in the end of the slide-plunger-chest escapes into the right-hand end of the main-cylinder, thereby breaking the equilibrium of the steam in the chest by removing some of the pressure from the right-hand end of the plunger; then the full steam-pressure acting on the left-hand end of the said plunger immediately forces it to the right, carrying the slide-valve with it. This movement of the slide-valve admits steam to the right-hand end of the main-cylinder, arrests the progress of the piston in that direction (by adding the steam load to the right-hand side of the *steam*-piston to the water load on the right-hand side of the *water*-piston), and at the same time opens the exhaust to the left-hand side of the piston and the end of the cylinder; the piston then performs its stroke from right to left, and the operation just described is carried out at the left-hand end of the cylinder.

'The parts, generally, are so accessible that, although still hot with steam, every detail may be examined in a few minutes. The steam and exhaust-pipes are connected to branches on the main-cylinder casting, so that the slide-chest, plunger, and valve may be at once taken off without any disturbance of pipes whatever.'

The pump-piston works in a cylinder which communicates at each end with two valve-chambers, each containing a suction and a delivery-valve (Holman's Patent). These valves are plain, light pieces of metal, without screws, pins, or joints, and having no *central* seatings, wings, arms, or grids to obstruct the passage of the water, they contribute considerably to the durability and good working of this important part of the machine. The action is very simple and is easily understood. We will suppose that the piston is moving from the right to the left. As it approaches the left-hand valve-chamber, it forces open the delivery-valve, and discharges the water contained in the chamber. On the return stroke the pressure being removed, the delivery-valve closes, and the suction-valve opens to fill the vacuum in the chamber. Approaching the opposite chamber, a similar action takes place. It is thus seen that the pump in common with others is double-acting.

'The pump-valves and seats are constructed of mineralised junction india-rubber, one portion of which is rendered hard for the purpose of keeping it true (in shape) and holding it in position, and the other portion is comparatively soft, to insure a silent and sound seating of the valve.

'The valve, instead of being a flat disc, is made of a parabolic form, so that there is less resistance to the passage of the water.

'The hard rubber seatings are secured in their places by screwed brass rings, affording every facility for examination or renewal when necessary. A tubular buffer (consisting of seven pieces of solid rubber, each containing about four cubic inches) is placed above each valve, the upper end embracing the guide-spindle, and the lower end the boss of the valve, thereby preventing any leakage through the centre of the valve, at the same time preventing too great a rise of the valve; moreover, it acts as a *spring*, assisting a speedy closing of the valves at the return stroke of the engine.

This is a most important feature, because the valves close automatically, instead of being hammered on to their seats by the rapid return of the piston under a heavy load of steam, which equals on a piston 12 inches diameter, at 40 lbs. per square inch, a weight of *two tons*. By the arrangement here shown we have a valve that is as simple, more effective, and much more durable than any clack-valve doing the same duty. They have worked for more than two years under heads of 500 feet without changing, and during the last seven years about 17,000 of them have been used in the "Special" steam-pump alone.

'These pumps are of various sizes, and at first only small ones were made, but as their usefulness became developed the manufacturers designed pumping engines on the same principle for use in collieries. They were first applied to this purpose in collieries about three years since (being two years after their introduction into this country). and through the efforts of the late Mr. Alfred Stansfield Rake, under the direction of the Messrs. Tangye, about 130 of these pumps had been introduced, principally in the collieries of the Durham and Newcastle districts, up to the end of 1870. They were adapted to perform the required duty, varying in almost every case, of forcing 1,000 to 10,000 gallons per hour from depths ranging from 100 to 500 feet. This led Mr. J. Bigland, the manager of Messrs. Pease's Bishop Auckland Collieries, to conclude it was adapted for yet heavier work. As the result of this an engine was contracted for having a 26-inch diameter steam-cylinder by 6½-inch water-cylinder, with a stroke of 6 feet. In the contract it was stipulated that the engine should raise 120 gallons per minute 1,040 feet high in a single lift, and this it accomplished with apparently as much ease as if its load was delivered at only 100 feet high. It was started on June 6, 1871, and Mr. Bigland reported that, having measured its duty, the average of seven trials was found to be 137 gallons per minute, thus exceeding the contract stipulation by 17 gallons per minute.*

'As I had not heard how this engine was continuing its

* *Engineering*, Sept. 6, 1872.

work for the last twelve months, I took occasion to write to Mr. Bigland for current practical data, to which he replied promptly on February 25, 1875, as follows:

'"In reply to your favour of the 23rd, the underground pumping engine at Adelaide Colliery is working night and day almost continuously, standing only 12 hours on Sundays. Some of the cupped leathers which form the plunger-packing have lasted three months, but a month may be taken as an average. The average duration of the valve-seats is about eight months; they work and keep tight while there is a bit of them left. It is not necessary to face them up unless a piece of coal or any other hard 'gag' gets in and spoils the face. If the injury is deep we melt the face and insert a piece of gutta-percha before putting the seat in the lathe. I expect the valve (lids) and the buffers will last as long as the colliery." This pump is fitted with multiple metal-valves and gutta-percha seats. There are 28 of these seats in the pump, and they can be renewed for 3*s*. 6*d*. each. The total cost of renewal of the valve details, under a load of 1,000 feet head, spread over two years, is fourteen guineas. Mr. Bigland adds for information: "I suppose you would be informed about the pump working without the air-vessel, and as far as we could see its removal did not make a bit of difference to the engine. Soon after the pump commenced to work (in 1871) one of the valve-boxes split (although tested to a head of 1,500 feet); at that time it was a *double-acting pump*, working single shift; and for several weeks, while new boxes were being made, we worked it as a *single-acting pump*, and kept it going double shifts."

'This was a very severe test, proving the complete command of the engine by the rapid movement of the slide-valve, by effecting the noiseless reverse of its strokes alternately—in one direction without a load, and in the other direction under a load of 1,040 feet head, plus the friction. I venture to submit that this very genuine and satisfactory report settles conclusively the great question of the practicability and durability of these direct-acting steam-pumping engines, both in their design and detail, for raising water from mines, however deep, in shafts or inclines of any line or angle, straight, obtuse, or acute; and in fact accomplishes in a unique and powerful machine Savery's dream of one and three-quarters century ago.'*

In the form of this pump specially adapted for high lifts the water-cylinder contains only one suction and one delivery-valve, and instead of a piston a differential ram is used.

The annexed sketch, fig. 16, represents the arrangement, the operation of which will be readily understood from the following explanation. When the ram is moved in the

FIG. 16.

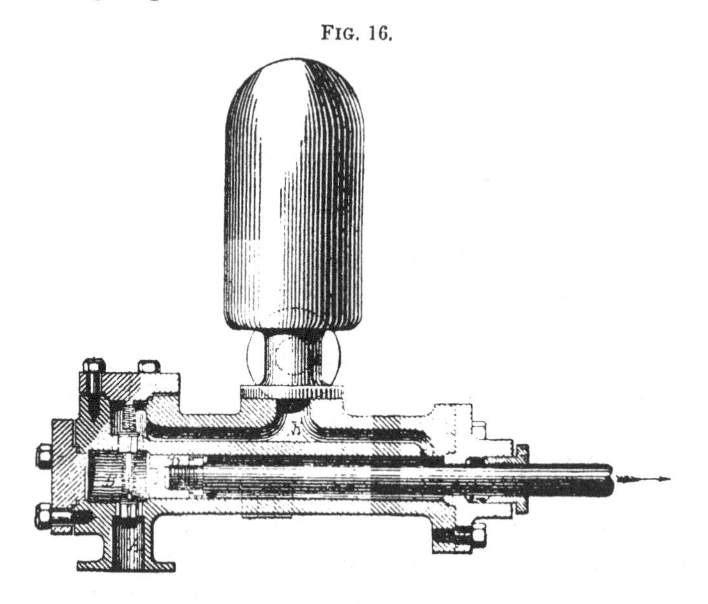

direction of the arrow F, the water enters the pump through the suction-valve (A), and a quantity of water equal to the difference of the areas between C and D (which should in all cases be one half, so that the work done at each stroke may be equal), is driven through the passage E into the rising-main. At the return stroke the suction-valve closes, and the D portion of the ram drives the

* 'History of Methods of Draining Mines by Non-rotating Steam Appliances.' By Mr. Stephen Holman. (*Trans. Chesterfield and Derbyshire Inst. Mining, Civil, and Mechanical Engineers*, April 1875.)

water through the delivery-valve B, part of which goes up the rising-main, and the other part refills the barrel around the smaller portion of the ram at C, hence the pump maintains a constant discharge. At the forward and backward strokes water is forced through the discharge-pipe, by which a continuous delivery is maintained, but it is only at the backward stroke that the inlet from the suction-pipe takes place, the valve being closed during the forward stroke. This pump was invented by Mr. Henry Thompson, late foreman to Sir William Armstrong, about the year 1851.

With regard to the durability of this pump we must refer the reader to the remarks a few pages back of the result of working the 26-inch pump at the Adelaide Colliery, for a period of two years, under above a thousand feet head. The gutta-percha beats for the valves last on an average eight months. At the time of reporting, February 1875, all the metallic portion of the valves were still in the pump, since June 1871—nearly four years—and were likely to last for many years. The exchange of a complete set of the seats can be effected in an hour; hard rubber is more durable than the gutta-percha. Mr. Bigland, of the colliery referred to, expects the valve lids and the buffers to last as long as the colliery.*

The reversing-valves or tappets at the ends of the cylinder have often been found a subject of frequent delay owing to their imperfect action, caused by accumulation of improper lubricants and the introduction of dirt, &c., due to want of proper care in working. Under such circumstances they are liable to stick, and sometimes require occasional cleaning, and at times they need to be replaced by new ones. Some may consider this an important objection. When in the Aberdare Valley the writer was informed by an engineer who had experience in the use of this pump that for dirty water the valve mechanism is too delicate; if the least dirt get between the tappets (reversing-valves) they do not get back to their places. Referring to a pump used at the West Yorkshire Colliery, pumping 465 feet, and working day and

* Holman's *History of Draining Mines*, p. 126.

night, it was reported that for twelve months the reversing-valves had been cleaned only six times; one bucket (leather) had been changed which had been worked since the pump was had (about two years), and one india-rubber clack-seat changed, representing as the cost of repairs for two years about £1. The attendance required to keep the pump running was one man twice a day, and that to replenish the tallow cups.* Another objection, and one which appears to be common to all or most pumps of this class, is the liability of the water-passages to get choked with coal dross, gravel, &c., when the latter are present in the water.

Fig. 14 shows Holman's Patent Condenser attached to this pump, which may be used with or without it. The condenser forms part of the suction-pipe, and whilst it effectually condenses the exhaust steam, it produces an average vacuum of about 10 lbs. per square inch on the steam-piston, and it is claimed that it effects a saving of from 20 to 50 per cent. in fuel, and increases to a corresponding extent the efficiency of the engine. It is suitable for any kind of steam-pump.

The following is a description of this condenser by the inventor, Mr. Stephen Holman, with a valuable account of its use in the drainage operations at Messrs. Lamb and Moore's Collieries at Wigan:

FIG. 17.

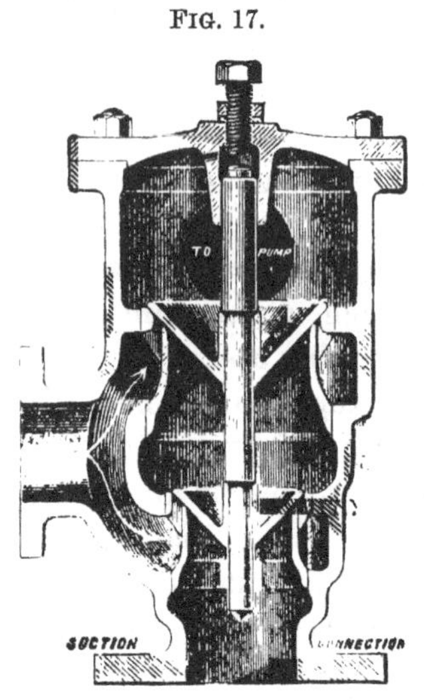

'The condenser, in its general structure, consists of a valve-box, having one or more valves (or deflectors) with single beats, and seats in which are formed annular steam spaces, through which the exhaust steam (at whatever pressure it leaves the cylinder) issues in thin annular streams to meet the water passing over the valve-seats.' Fig. 17 is a section of the condenser.

* Holman's *History of Draining Mines*, pp. 127 and 128.

'In the conducting-pipe for conveying exhaust-steam from the cylinder to the condenser is fixed a novel arrangement termed a "blow-through" valve, represented in the diagram below, fig. 18, which takes the place of two ordinary stop-valves. The disc-valve at the bottom of the spindle is faced on both sides, and when screwed up closes the outlet to the atmosphere, thus carrying the steam in the direction of the lower arrow towards the condenser. When the pump is first started, it is necessary to exhaust into the atmosphere, in order that the pump, cylinder, and pipes may be fully charged with water before the steam is let in; and this is effected by screwing down the disc, thus closing the way to the condenser, and opening a passage to the atmosphere in the direction of the upper arrow.

FIG. 18.

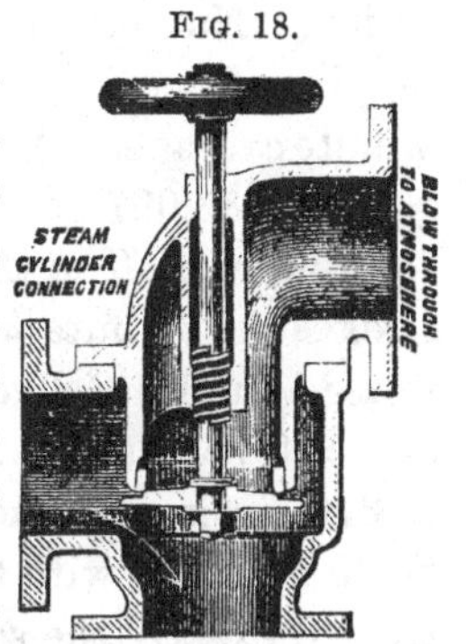

'The useful effect will greatly depend upon the proportion of the steam and pump-cylinders. Where the former is largely in excess of the latter, the vacuum given on the larger area will be multiplied in its effect upon the other, just as in a steam-cylinder 12-inch diameter we have four times the area of a pump-cylinder 6-inch diameter, so 10 lbs. of steam in the former will be in effect equal to 40 lbs. on the latter. The results due to the condensation of the exhaust steam are practically affected by ever-varying conditions of work. The following is a report of tests made at Messrs. Lamb and Moore's Collieries, Wigan, by Mr. Scarborough, with a pair of 30-inch × 10-inch × 48-inch 'Special' steam-pumps, capable of raising 40,000 gallons per hour 500 feet high in a single lift (under date August 17, 1875) at a piston speed of 100 feet per minute:—

'These engines are raising the whole of our water in from six to seven hours, or, I may say, we collect our whole 24 hours' make of water in a lodge in the workings, from which the engines pump it to bank at night after we have done winding coal. They are raising, when at work, from 40,000 to 50,000 gallons per hour, performing from 14 to 17 strokes per minute, and have Holman's Patent Condensers attached. Our practice, previous to putting down the pumping engines, was to wind the water at night at the two winding shafts, and we cleared by this process our whole 24 hours' make of water in from 11 to 12 hours.

'I have been induced to make a number of careful tests with a view to satisfy myself as to the value of your condensers as economisers of steam, and as to their action, if any, in relation to the quantity of water actually delivered to bank. I will describe the tests I have made for quantity of water actually raised to bank with the condensers in action. Number of strokes performed, at various speeds, ranging from 12 to 25 strokes per minute $=370=18\frac{1}{2}$ per minute average. Actual quantity of water delivered $=17,699$ gallons $=53,097$ gallons per hour (the test for quantity being 20 minutes). The theoretical quantity of water due at $18\frac{1}{2}$ strokes per minute $=57,651$ gallons per hour. Quantity of water actually delivered, compared with that theoretically due $=92$ per cent. The tests for quantity of water thrown in 370 strokes, at the same varying speeds, during a period of 20 minutes, with the steam exhausting to the atmosphere, yielded exactly the same result, as nearly as I could measure the water. I may here state that, in measuring the quantity of water delivered, I made use, in all cases, of a water-tight lodge, or pond, into which the water is discharged from the delivery-pipe of the pumps; the height of water being indicated by a scale of inches. The superficial contents of this lodge range from 390 to 483 gallons per vertical inch of water, by scale, and the height to which this lodge is filled in a given time, from a given mark on the scale, I take to be a correct index of the quantity delivered. Having satisfied myself that the quantity of water thrown per given number of strokes was the same, with the condenser on, or exhausting to the atmosphere, the testing of economy was a short and easy task. With the steam supply regulated to produce 12 strokes per minute when exhausting to the atmosphere, 18 strokes per minute were obtained when exhausting to the condensers: thus showing an economy of 50 per cent. With the steam supply regulated to produce a speed of 16 strokes per minute when exhausting to the atmosphere, 22 strokes were obtained by exhausting to the condensers: showing an economy of 38·75 at this speed. These results I have confirmed by several subsequent tests, and I make but little comment on them. If they had been against you, they would certainly have been as plainly stated. They are eminently satisfactory to us, as I do not doubt they are to you, and are thoroughly conclusive as to the advantage gained by the system of condensation. Relative to your inquiry as to the result of our experience of the system of pumping as a whole, I may say that it is really practically economical; we have proved it to be much more so than the system of winding water, even in the actual use of steam; and the saving of labour at the two winding shafts will, in our case, pay for fully 33 per cent. of the quantity of coal consumed in pumping the whole water.'

An addendum to the foregoing report was given, under date August 27 last, as follows:—'As to actual pressures of steam in the cylinders, with the condenser in and out of action, when the engines are running 18 strokes per minute, exhausting to the atmosphere, the steam pressure in the cylinder is 31½ lbs. per square inch by gauge; at the same speed, exhausting to the condenser, the steam pressure in the cylinder is 21 lbs. per square inch. When the engines run 22 strokes per minute, exhausting to the atmosphere, the steam pressure in the cylinders is 36 lbs. per square inch; at the same speed, when exhausting to the condenser, the steam pressure in the cylinder is 25 lbs. per square inch. Thus we have a corresponding gain indicated by the reduction in the pressure of steam required when exhausting into the condenser, to the increase of speed immediately given by turning the exhaust steam from the atmosphere to the condenser.' *

* Extracted from the *Royal Cornwall Polytechnic Society's Report for* 1875, pp. 76–80.

THE 'UNIVERSAL' STEAM-PUMP

Like the 'Blake' and 'Special' and other valuable engines of a like kind, this pump was invented in the United States, and has for its authors, Messrs. Ezra Cope and James Riley Maxwell, of Cincinnati. It was patented in England, September 15, 1868, since which time a great many improvements have been effected in it, and many thousands have been made by the well-known firm of Messrs. Hayward Tyler and Co., the sole makers for this country.

There are two types of this engine—the long piston, short-stroke engine, and the short piston, long-stroke engine, each of which has two forms of pumps. It will therefore be necessary to consider the following four kinds of the 'Universal' steam-pump, viz.:

I. *The Short-Stroke Engine with a Long Piston.*

- *a.* With a 'Universal' double-acting piston-pump.
- *b.* With a double-acting ram-pump.

II. *The Long-Stroke Engine with a Short (i.e. an ordinary) Piston.*

- *a.* With a 'Universal' double-acting piston-pump.
- *b.* With a double-acting ram-pump.

I. The Short-Stroke 'Universal' Pump.

a. With a Double-acting Piston-Pump.

This engine is represented by figs. 19 and 20, and in section by figs. 21 to 26. It will be observed from figs. 19 and 20 that there are two forms of water-chamber for this engine; that shown in fig. 19 is used with engines of 4-inch to 9-inch steam-cylinders and 2-inch to 7-inch pumps, and that shown in fig. 20 is used with engines of 9-inch to 21-inch steam-cylinders and 10-inch to 15-inch pumps. It is used for lifts not exceeding 200 feet, and where very rapid action is desirable, as, for instance, for emptying wrecks, docks, &c., draining flooded mine-workings, and for any other purpose where the lift is comparatively low, the instantaneous reversing action of

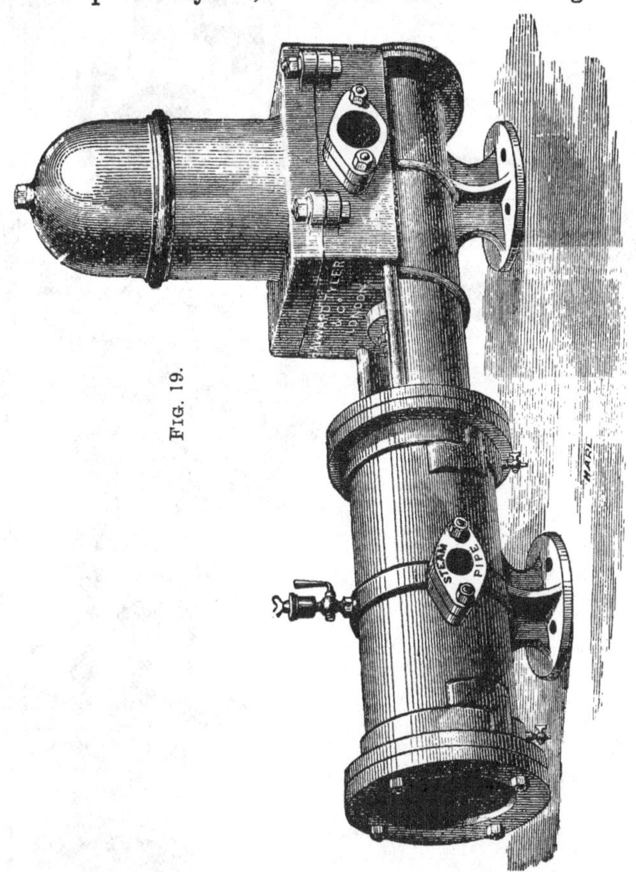

Fig. 19.

this pump gives it advantages which do not appear to be possessed by any other pump. It is largely and very successfully applied to raising sunken or sinking ships for salvage purposes. Pumps of 15-inch steam and water-cylinders, delivering from 70,000 to 80,000 gallons an hour, occupying a space of 36 square feet, have been used for work of this kind.*

* *Direct-Acting Pumping Machinery*, by Mr. J. C. Fell.

The illustrations referred to show that this pump has a very long steam-cylinder, and that, unlike all other machines of this class, it has no steam-valve-chest nor any appearance of

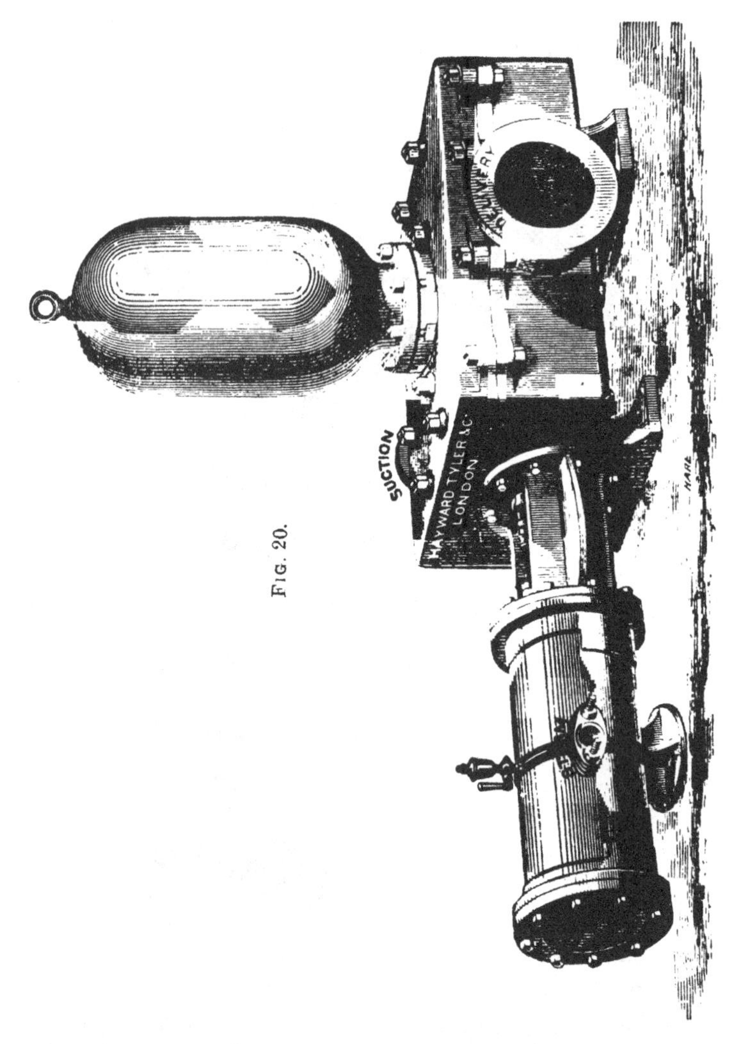

Fig. 20.

valve-gear external to the cylinder, but that in lieu thereof it contains a unique piston of considerable length, within which is enclosed the valve necessary for imparting motion to the engine. Fig. 21 is a sectional elevation of the steam and water-

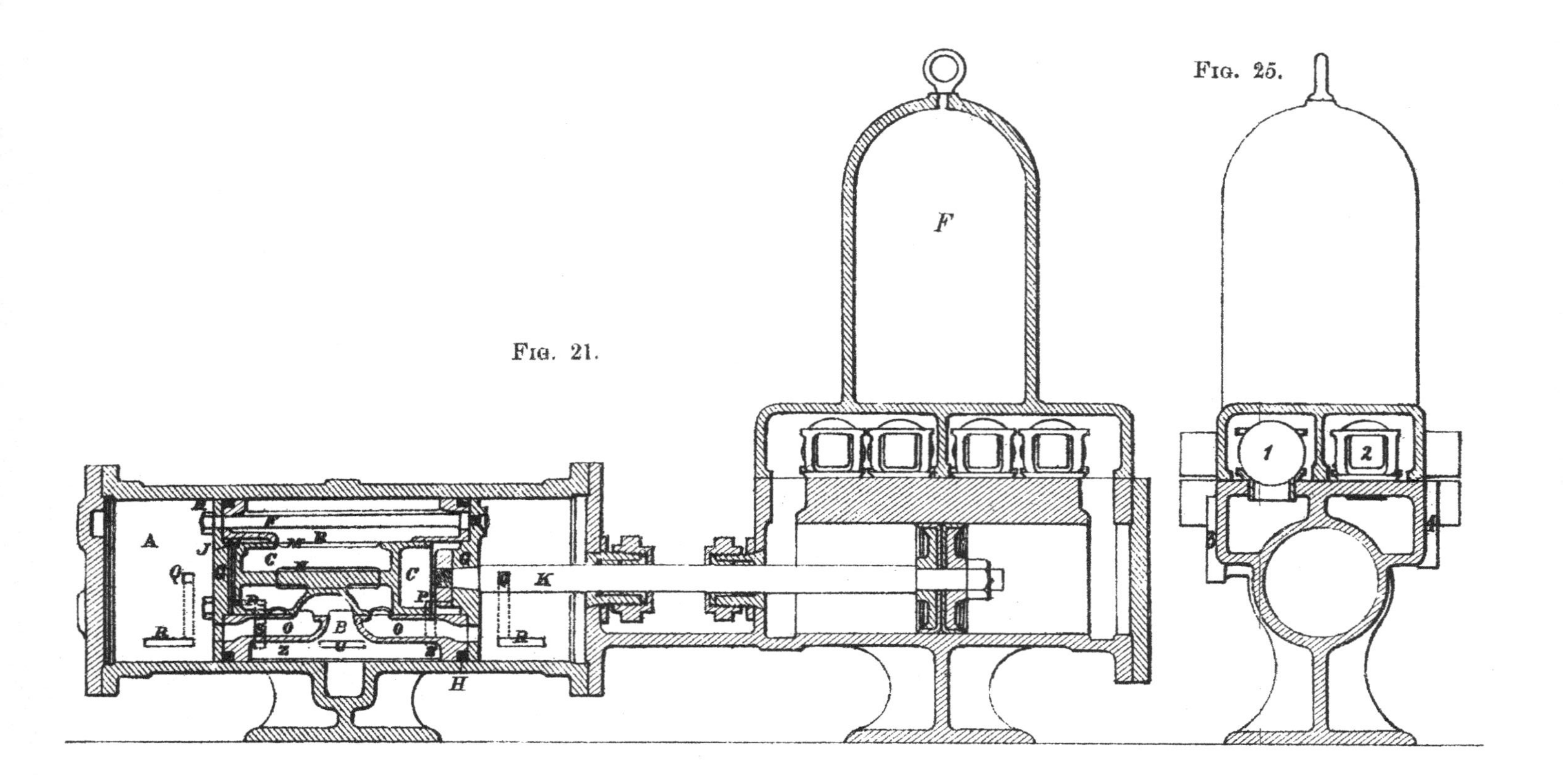

Fig. 21.

Fig. 25.

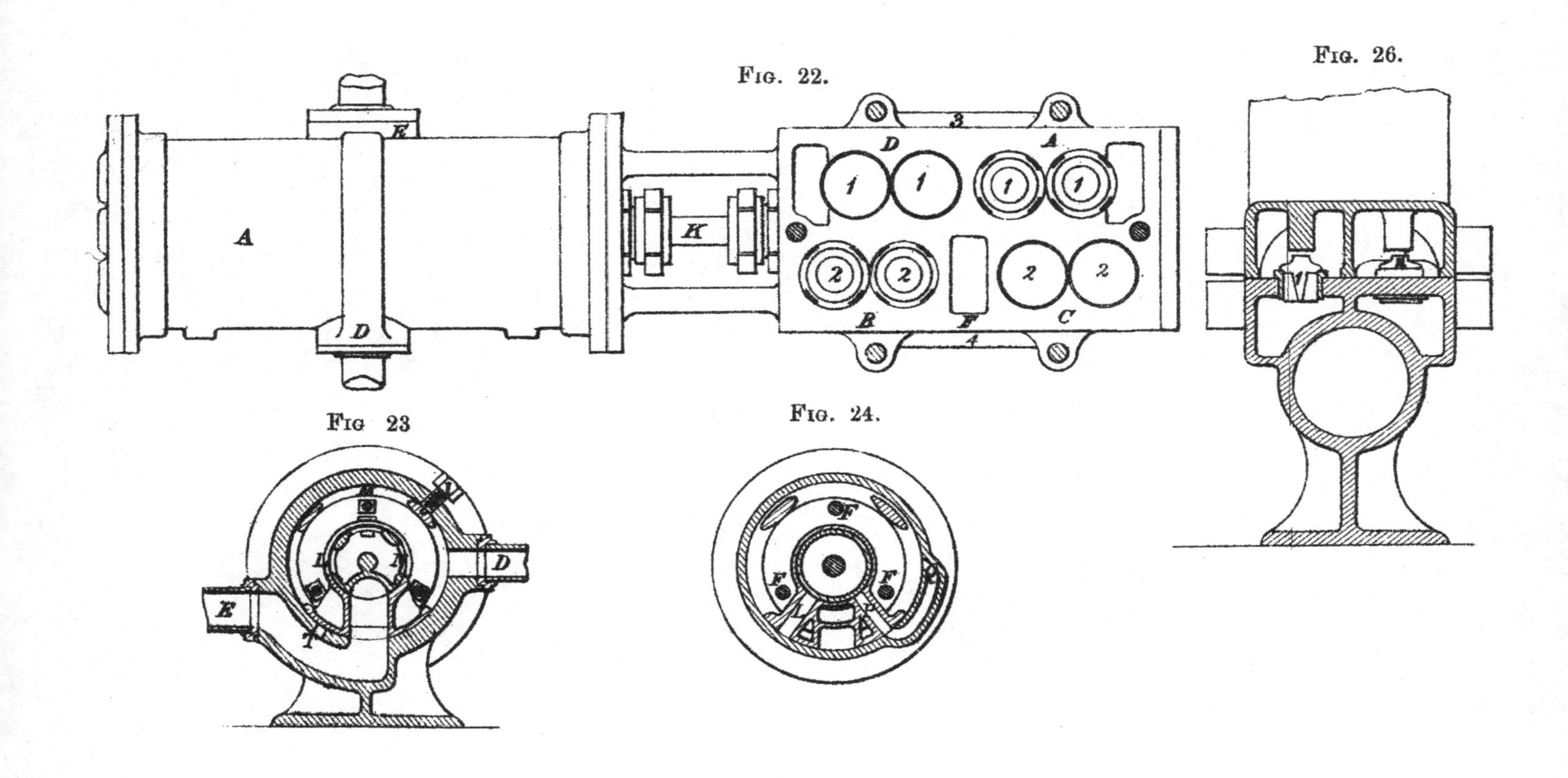

Fig. 22.

Fig 23

Fig. 24.

Fig. 26.

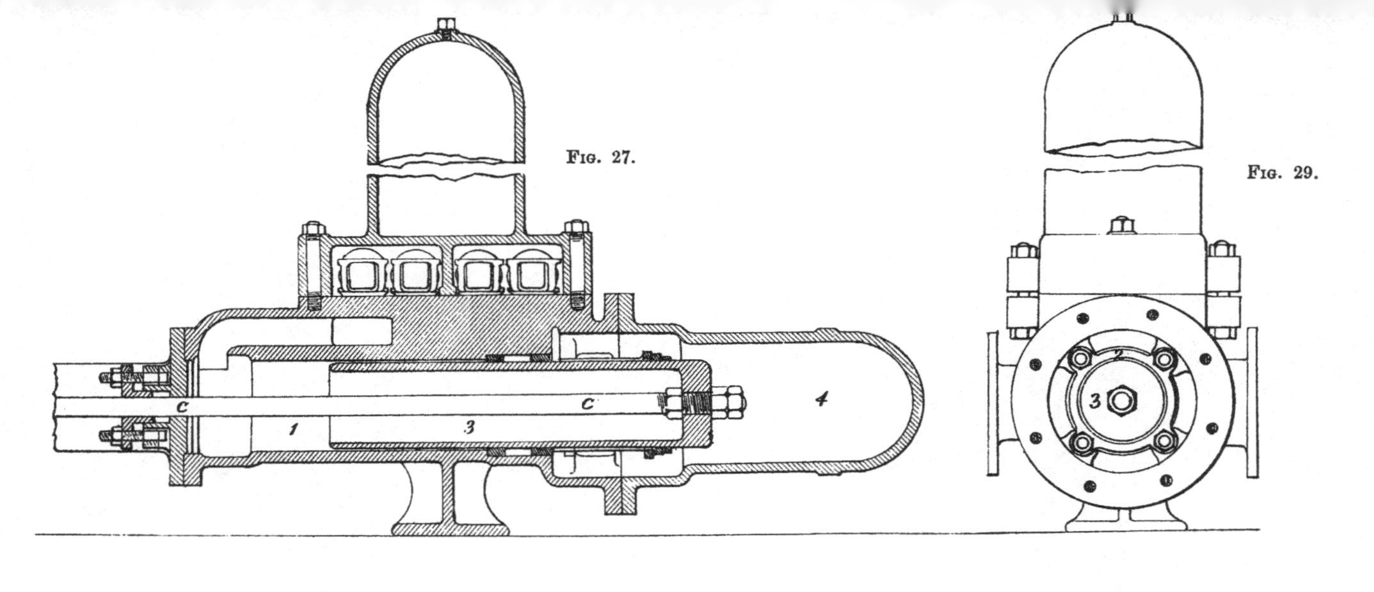

Fig. 27.

Fig. 29.

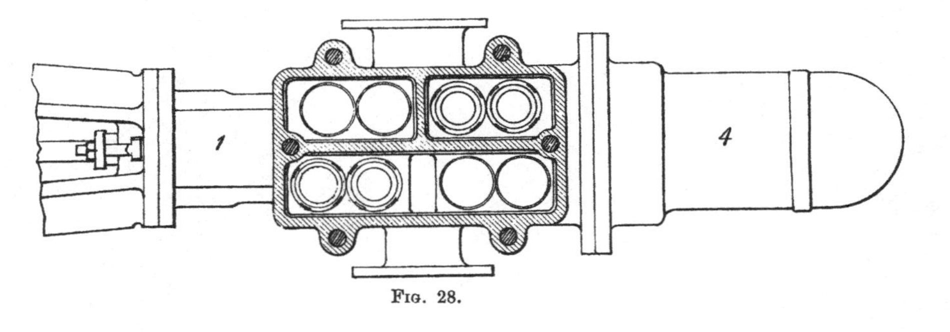

Fig. 28.

cylinders and their contents, and gives a very good idea of this machine; fig. 22 is a plan showing one group of suction (D 11) and one of delivery-valves (C 22) removed; 3 is the suction- and 4 the delivery-pipes; fig. 23 a section through the exhaust-port; fig. 24 a section through the shooting or valve-moving ports; fig. 25 a transverse section of the water-cylinder and valve-chamber, showing a section and elevation of the india-rubber ball-valves and seats; fig. 26 is a section through the same place showing ordinary gun-metal mitre-valves. India-rubber disc-valves were formerly used, but have been discarded in favour of the ball-valves, which are available for heads up to 200 feet. For greater heads the gun-metal mitre-valves are used. As the action of this engine is rather difficult to understand, a minute description is given of the details and the irfunctions, in order that the patient reader may be able to follow the movements of the different parts as afterwards explained.

The steam-cylinder is a plain barrel with a steam-branch, D (fig. 22), on one side, midway between the two ends, and an exhaust-branch, E, on the other side and opposite, which communicates with the bottom of the cylinder. Near each end of the cylinder on the same side, and midway between the top and bottom, is a steam shooting-port, Q (fig. 21), communicating by a passage with a long horizontal slot, R, admitting steam to one end of the valve-chamber. Midway between these slots, but on the opposite side of the cylinder, is the exhaust-slot. T, communicating alternately with each end of the valve-chamber by corresponding exhaust-ports on the opposite side. The piston, B, is packed at each end with ordinary packing-rings. The covers of the piston are held together by bolts. A guide-pin, I, screwed into the cylinder and fitting a corresponding groove in the piston, prevents the latter from rotating. The length of the piston between the inner side of the ends must be equal to the length of the stroke in addition to the steam-inlet. The body or outer shell of the piston between the two ends is quite open at the top and sides except where two comparatively narrow distant-pieces near the top run the whole length of the piston. The bottom part, in which are two steam-ports, O O, and the exhaust-port, Z, works steam-tight on the cylinder. In the centre of the piston body is a cylindrical valve-chamber in which a long piston, C, constituting the slide-valve, works. A guide-pin, J, in the valve, which works in a groove in the valve-chamber, prevents it from turning. We will now proceed to describe the action of the valve and piston. The steam has admission to both ends of the piston alternately through one of the ports, O, which causes it to make a stroke towards the opposite end. Near the termination of the stroke, the piston carries the port, S P, over the long slot, R, which communicates with the aperture, Q, and Q being in a horizontal line with the open part of the piston filled with live steam from the boiler, immediately the end of the piston has passed over Q steam is admitted through Q, R, S, P, to the back of the piston-valve, the slot, R, being made sufficiently long to admit to S P the whole of the steam received at Q throughout the termination of the stroke. The steam thus entering at S P acts on the back of the valve, and the port, L (fig. 24), at the opposite end being simultaneously in communication with the exhaust-slot, T (fig. 23), causes the valve to make a stroke in the opposite direction, and, thus admitting steam to the other end of the cylinder, causes the piston to make the return stroke.

We now begin to describe the water-end of the pump. This consists of a cylinder with a valve-chest mounted upon it. The latter is divided into three compartments—two for two pairs of suction-valves, and the other running the whole length for the four delivery-valves. The top part of the valve-chamber with the air-vessel is a separate casting, which is bolted to the cylinder and base for the valve seatings, which are another casting. The valves may be india-rubber balls, or brass, or gun-metal wing-valves.

The action of the pump is as follows. As the piston approaches the left-hand end of the cylinder, the water enters through the suction-valves in the chamber, A, and is discharged through the delivery-valves, 2. 2. B. On the return stroke the water that entered through 1. 1. A at the previous stroke is forced through the delivery-valves, 2. 2. C; meanwhile water enters through the suction-valves 1. 1. D. The whole of the water is discharged through the same outlet, F (fig. 22), over which stands the air-vessel F (fig. 21).

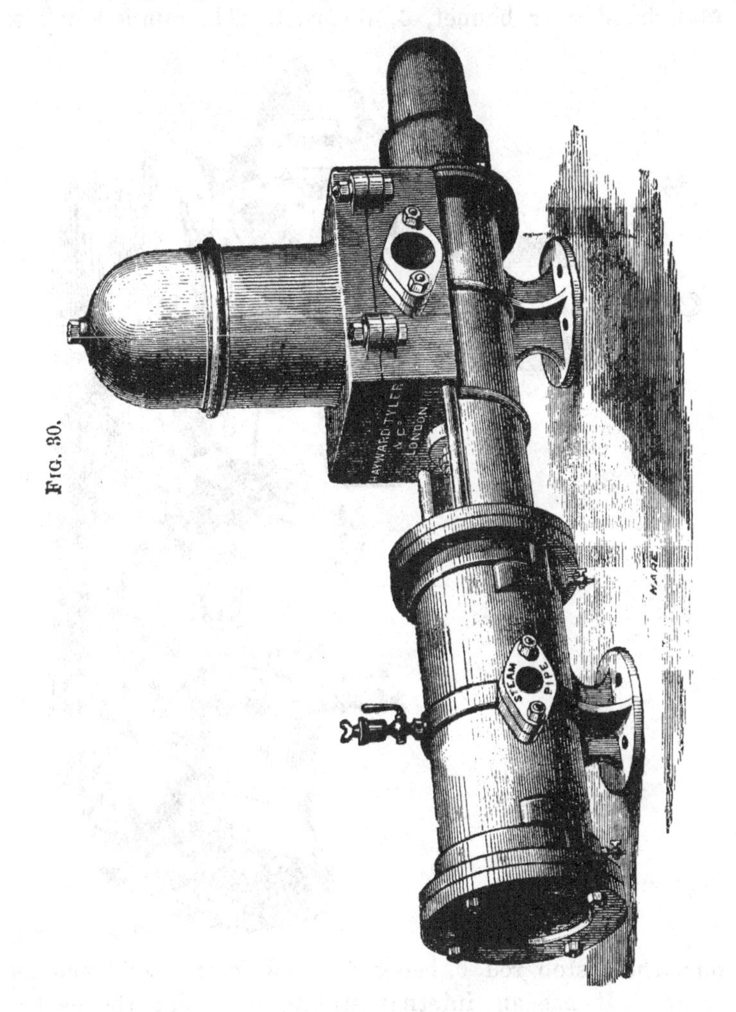

Fig. 30.

b. With a Double-Acting Ram-Pump.

This adaptation of the 'Universal' engine just described to a ram-pump is shown at fig. 27, which represents a longitudinal section of the water-cylinder, valve-chambers, ram, &c. Fig. 28 is a plan, with the valve-chambers in section, the air-vessel and valve-chamber-cover, cast in one, being removed, and fig. 29 is an end elevation, with the ram-chamber or bonnet, 4, removed. The ram is a hollow

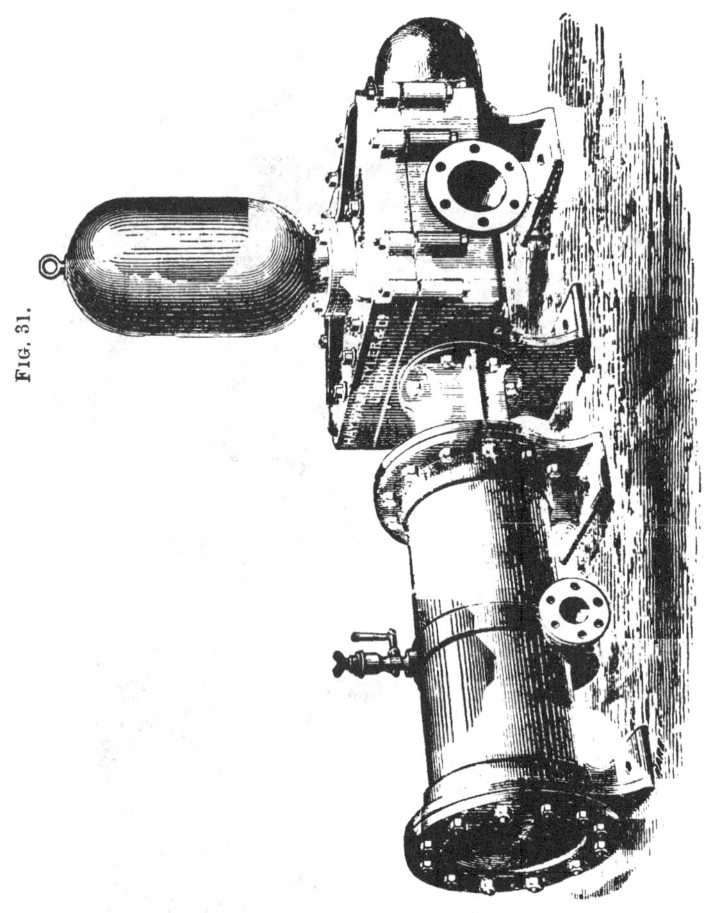

Fig. 31.

one, the piston rod, c, being screwed to the solid end, as shown. It has an internal stuffing-box near the centre (2, fig. 29), to get at which the bonnet, 4, must be removed. The same kinds of valves are used with this pump as with the piston-pump.

There are also two kinds of this pump—one for engines

of from 5 to 9-inch cylinders and 3 to 10-inch pumps, and one for engines of from 12 to 21-inch cylinders and 6 to 12-inch pumps, of which figs. 30 and 31 are perspective elevations respectively.

This pump is used for heads not exceeding 200 feet; and for mines and collieries where the water contains more or less solid impurities, it is preferred to the piston-pump.

A combination of the short-stroke engine and a ram with external stuffing-boxes, shown at fig. 32, was formerly

Fig. 32.

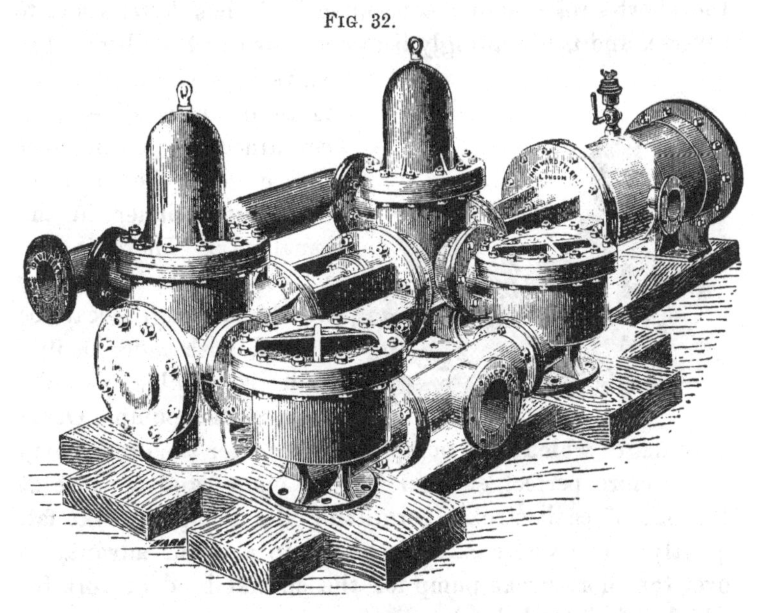

used for high lifts, up to 700 feet, but has been abandoned in favour of the long-stroke engine with the same form of pump.

II. The Long-Stroke 'Universal' Pump.

a. With a 'Universal' Double-Acting Piston-Pump.

For very high lifts, where a slow motion is indispensable, and for general mining purposes, this pump is a superior machine to the short-stroke pump.

Having fewer times to reverse, and consequently fewer times to fill the steam-ports and clearances, than the short-stroke engine, to give the same piston speed it would be natural to think that this engine is a more economical one than the short-stroke engine; but the author has the authority of Messrs. Hayward Tyler and Co. for stating that such is not the case, experience having proved that the unique long-piston engine takes less steam than the long-stroke engine; the steam being in the piston, there are no long ports to fill. The steam-piston is of the ordinary form, and permits of a stroke more than twice the length of that of the short-stroke pump; consequently, it has fewer times to reverse, and is accordingly better adapted to high lifts. The valve is steam-moved, and is an ordinary slide-valve, contained in a chamber placed on the top of the main-cylinder in the usual way. There is also a lever attached to it, by which the engine can be started with more facility than is possible with the short-stroke pump. Other advantages which this pump possesses over the other pump are a more perfect cushion for the piston and a pause at the end of each stroke, permitting the pump-valves to fall quietly. The water part of the pump also has advantages over the short-stroke pump for the special kind of work for which it is intended.

FIG. 33.

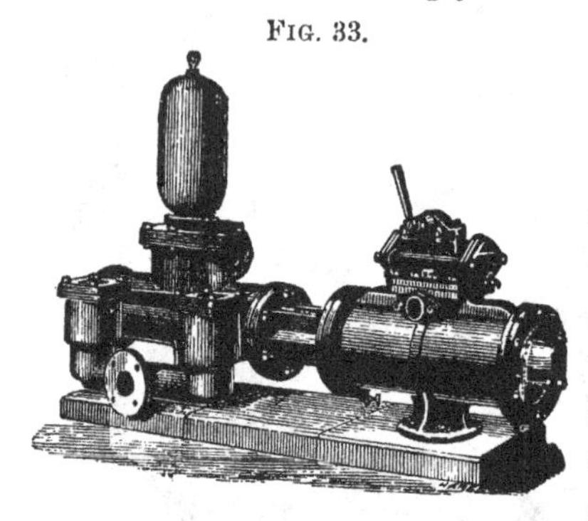

The steam-cylinder, valve-chest, &c., are shown at fig. 33, and the larger figs. 34 and 35, and the pumping part is a modification of that shown in connection with the piston-pump with the short-stroke engine at fig. 21.

The valve-chest, as before remarked, is mounted upon the steam-cylinder in the ordinary way, and contains a main piston-valve, in the interior of which, and near the top thereof, is a small 'leading' or 'shooting'-valve, which is acted upon by steam through one of the valve-moving ports, the other end of the leading valve being simultaneously in communication with one of the exhaust-ports. This causes it to travel to the opposite end of its chamber, in doing which it opens steam and exhaust communication to each end of the main valve-chamber alternately, causing the main-valve to make a movement identical with the leading valve, thereby changing the steam and exhaust communications with each end of the main steam-cylinder alternately. The cylinder is a plain barrel with two main steam-ports, and two small cushioning-ports, one in each end of the cylinder at the top. It also contains the valve shooting-ports. The piston is of the ordinary type, with the usual packing-rings.

The main steam-ports are kept back from the ends of the cylinder, and small cushioning steam-ports only are in communication with the extreme ends, so that the exhaust may be partially shut off before the termination of the stroke, and the steam admitted gradually at the return stroke to prevent 'knocking' in the pump-valves, giving them more time to settle in their places. The pump can be started or stopped in a moment by a small hand-lever which is fixed on the top of the main valve-chamber, and which actuates a quadrant by means of a spindle working through a stuffing-box.

b. With a Double-Acting Ram-Pump.

Of the various forms of pumps which Messrs. Hayward Tyler and Co. manufacture, the above is *par excellence* the pump for mine and colliery use, particularly for heavy lifts, and may be used for heads up to 1,000 feet. Figs. 34 and 35 are an elevation and plan of this well-known engine, and figs. 72 to 75, which are views of the self-governing engine, show the arrangement of the pump—the same kind of pump being used with both engines.

It will be observed that this engine has a long hollow ram working into two ram-cases and fitted with two outside stuffing-boxes, as all underground engines should be. The suction and delivery-valve boxes are shown at 4.4 and 5.5 respectively; 1.1 are the ram-cases and 2.2 the stuffing-boxes. We must refer the reader to pages 40, 41, and 79, for examples of work done by 'Universal' engines in collieries.

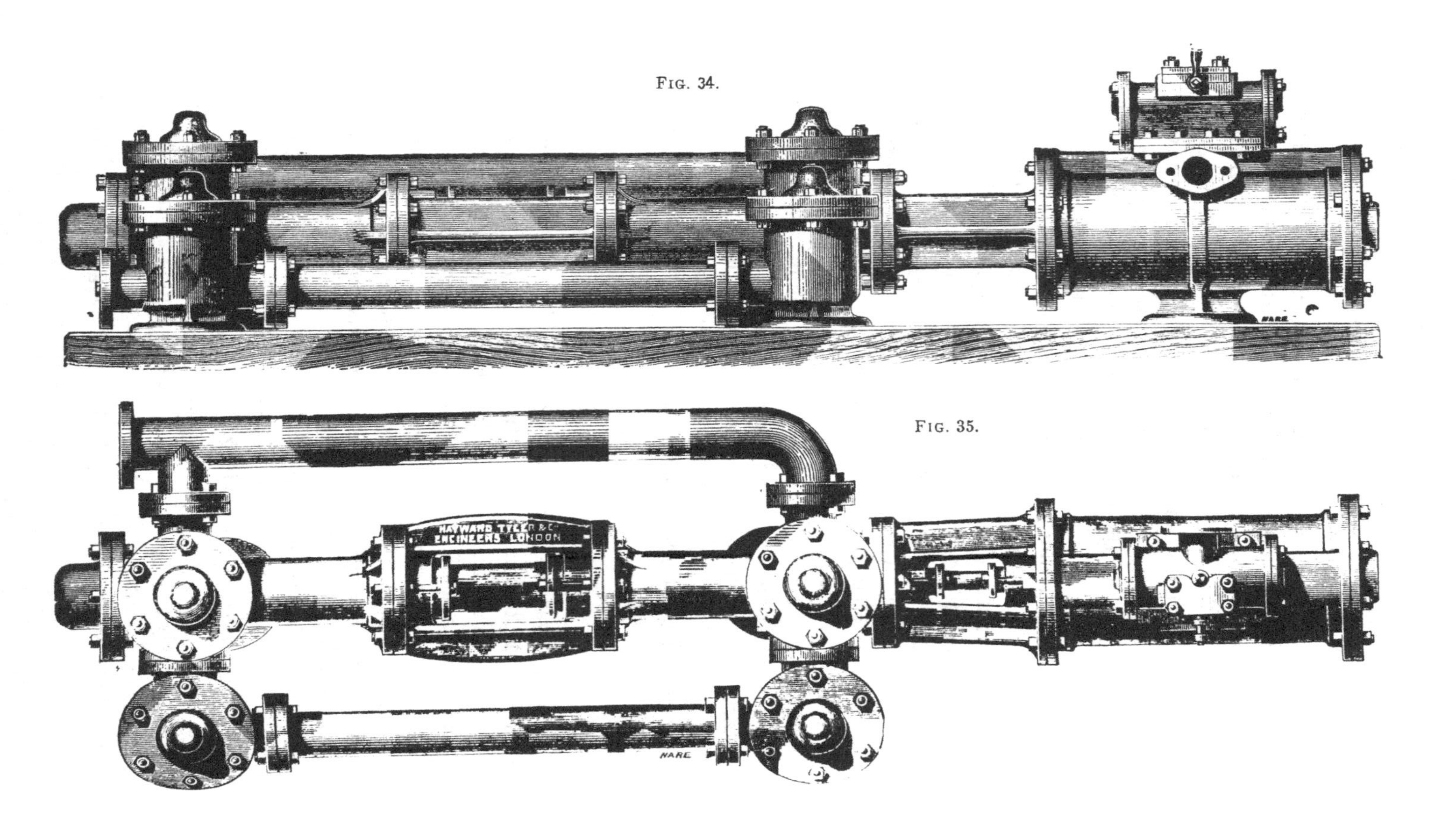

Fig. 34.

Fig. 35.

Messrs. Hayward Tyler and Co.'s list, which we give below for the useful information it contains relative to this engine, enumerates sizes varying from a 12-inch steam-cylinder and 4-inch ram to a 33-inch steam-cylinder and a 12-inch ram, but the largest size made in this form has been, we believe, a 21-inch steam-cylinder and 9-inch ram; though they have made their long-piston short-stroke engines with 33-inch cylinders, and several on a rather different plan with cylinders of 40 inches and rams of 10 inches diameter.

Diameter of steam-cylinder	Diameter of pump-plunger	Length of stroke	Diameter of suction-pipe	Diameter of delivery-pipe	Diameter of steam-pipe	Diameter of exhaust-pipe	Gallons per hour	Height in feet with 40 lbs. steam pressure	Strokes per minute	Total length	Total width	Weight (approximate)
in	in.	ft. in.	in.	in	in	in.	approx.	feet		ft. in	ft. in.	cwt. qr. lb.
12	4	2 0	2½	2½	2½	3	3,000	450	25	11 6	3 6	26 2 0
12	5½	2 0	3	3	2½	3	5,400	240	30	11 6	3 8	32 2 0
12	6	2 0	4	4	2½	3	7,000	200	30	11 6	4 0	40 0 0
15	4	3 0	2½	2½	2½	3	3,000	700	17	14 6	3 6	36 1 0
15	4½	3 0	3	3	2½	3	3,500	540	18	14 6	3 8	44 2 0
15	5	3 0	3	3	2½	3	4,700	430	18	14 6	3 8	44 2 0
15	5½	3 0	4	4	2½	3	5,400	360	20	14 6	4 0	45 0 0
15	6	3 0	4	4	2½	3	7,000	300	20	14 6	4 0	48 0 0
15	6½	3 0	5	5	2½	3	8,500	250	22	14 6	4 6	52 1 0
15	7	3 0	5	5	2½	3	10,000	220	25	14 6	4 6	52 1 0
15	7½	3 0	5	5	2½	3	11,000	200	25	14 6	4 9	55 0 0
18	4	3 0	3	3	3	3½	3,000	960	16	14 9	3 8	58 0 0
18	4½	3 0	3	3	3	3½	3,500	750	16	14 9	3 8	58 0 0
18	5	3 0	3	3	3	3½	4,700	600	16	14 9	3 8	58 0 0
18	5½	3 0	4	4	3	3½	5,400	480	17	14 9	4 0	62 0 0
18	6	3 0	4	4	3	3½	7,000	430	17	14 9	4 0	62 0 0
18	6½	3 0	5	5	3	3½	8,500	370	18	14 9	4 6	64 0 0
18	7	3 0	5	5	3	3½	10,000	310	18	14 9	4 6	64 0 0
18	7½	3 0	5	5	3	3½	11,000	270	20	14 9	4 9	68 0 0
18	8	3 0	5	5	3	3½	12,500	240	20	14 9	4 9	68 0 0
21	5	4 0	3	3	3½	4¼	4,700	800	13	17 6	4 0	88 0 0
21	5½	4 0	4	4	3½	4½	5,400	680	14	17 6	4 0	92 0 0
21	6	4 0	4	4	3½	4½	7,000	580	15	17 6	4 6	92 0 0
21	6½	4 0	5	5	3½	4½	8,500	490	16	17 6	4 6	94 0 0
21	7	4 0	5	5	3½	4½	10,000	430	17	17 6	4 6	94 0 0
21	8	4 0	6	6	3½	4½	12,500	320	18	17 6	4 9	98 0 0
21	9	4 0	6	6	3½	4½	16,000	256	20	17 6	5 0	103 0 0
26	6	5 0	4	4	3½	4½	7,000	864	10	20 6	4 6	101 0 0
26	6½	5 0	5	5	3½	4½	8,500	770	10	20 6	4 6	105 0 0
26	7	5 0	5	5	3½	4½	10,000	672	11	20 6	4 6	105 0 0
26	8	5 0	6	6	3½	4½	12,500	480	12	20 6	4 9	107 3 0
26	9	5 0	6	6	3½	4½	16,000	400	12	20 6	5 0	110 0 0
26	10	5 0	8	8	3½	4½	19,800	330	13	20 6	5 6	162 1 0
26	11	5 0	8	8	3½	4½	23,500	264	13	20 6	6 0	—
26	12	5 0	10	10	3½	4½	30,000	240	14	20 6	6 0	—
33	11	6 0	8	8	5	6½	23,500	420	9	22 6	6 0	—
33	12	6 0	10	10	5	6½	30,000	350	10	22 6	6 0	280 0 0

Referring to the different types of Messrs. Hayward Tyler and Co.'s engines for underground drainage, Mr Fell, in his paper on 'Direct-acting Steam-Pumping Machinery,'* remarks as follows:—'It may be interesting to detail some of the most suitable applications of these pumps (the short and long-stroke types). They have been found most exceptionally efficient under the most extreme diversity of lifts, from 10 feet to 600 feet, but it must be remembered that the conditions of success for such different pressures depend upon very widely diverse data. Under the low lifts, up to 20 or 30 feet, the most elastic class of pump-valves must be used, so as to ensure the least noise and wear and tear under high piston speeds. The lift of these valves must also be low, and should be distributed over a large space. This requirement has been admirably met by Messrs. Hayward Tyler's arrangement of multiple india-rubber ball-valves, of which a specimen, which has been subjected to many months' wear, is laid before you. It will be found that this valve has actually lost one-eighth of an inch in wear all over the ball so uniformly that its shape is not altered, nor is it marked in any way.

'Under these low lifts as much as 230 to 300 feet piston speed can be obtained from Hayward Tyler's "Universal" pumps, so elastic and lively is their action. Amongst the higher lifts, up to 60 or 70 feet, may be reckoned most of the applications to the ordinary factory, household, brewers', and small water-supply purposes, and up to about 150 or 200 feet the same multiple india-rubber valve system is most successfully used.

'Ranging upwards from these lifts to 600 feet vertical and more are the heavy colliery lifts. For these higher lifts a very different form or construction must be adopted. The piston speed must be much reduced and can seldom run more than 100 to 120 feet per minute. The india-rubber valves are no longer serviceable, at any rate in their

* Read before the South Staffordshire and East Worcestershire Institute, February 19, 1877.

original ball form, and either hard rubber-discs * or gun-metal valves must be used. In each case the lift of the valves must be very small, to reduce the shock of the reversal of the stroke, and for that reason also the reversals should be as few as possible, as long strokes are advisable for high lifts. The system of a double plunger-pump is most approved for the deep lifts, in which the plunger is made tight by two outside packing-glands, packed with ordinary hemp.

'Out of many testimonials of their efficiency, under such circumstances the writer need only mention one or two, such as a 20-inch cylinder with a 6-inch double-acting plunger-pump throwing 3,500 gallons per hour a height of 400 feet vertical, at a slow speed, in the Foxhole Collieries of the Patent Anthracite Coke Company; also at the Yate Collieries a 15-inch cylinder and 4-inch pump, throwing 2,500 gallons per hour, at a slow speed, to a height of 480 feet.' †

Before proceeding with descriptions of other engines, the author would conclude this account of Messrs. Hayward Tyler and Co.'s engines with a description of the valve-gear which has been patented by Mr. R. L. Howard, senior partner of the firm, and which is applicable to large engines of the 'Universal' long-stroke class, as well as to compound engines, in the same way as the valve-gears of Messrs. Cope and Maxwell and Mr. Henry Davey. The following is taken from the specification filed in the Patent Office:

'Heretofore the movement of direct-acting steam-engines has in various ways been controlled or regulated according to the varying load thereon by means of valve-gear, wherein the main-valve is worked by an auxiliary steam-cylinder controlled by a cataract or other governor,

* The rubber-disc valves have been discontinued, being inadequate for high lifts.

† The number of strokes per minute was 30, the steam-pressure 30 lbs. per square-inch, and the distance of the boilers from the pump 340 yards. A further testimonial dated January 14, 1880, after five years' work, states thus :—' We have never experienced any break-down, although the pump has been in constant work during the whole five years, and only a very small sum has been expended in repairs.'

and receiving a secondary motion from some moving part of the engine or by an expansion-valve working in an opposite direction to the main slide-valve, and controlled indirectly by a cataract.

'Now the object or purpose of this invention is to control or regulate direct-acting steam-engines according to the varying load by means of valve-gear wherein the expansion-valve is worked from the main steam-piston, the main-valve being worked by a subsidiary-piston controlled by a cataract or other governing appliance, all operating as hereinafter described.

'For this purpose I employ by preference an ordinary D-shaped main slide-valve, with ports passing through it, and I work an expansion-valve on it. The main slide-valve is moved by means of an auxiliary steam-piston, controlled by a cataract in a well-known manner. The expansion slide-valve is worked from the main steam-piston in the same direction and in unison therewith, but with a shorter stroke, the necessary connection being by a system of levers and connecting-rods, or any other well-known device for imparting the motion to the expansion-valve, the latter being moved in unison with the main-piston for the purpose in view, and operating in such a manner that when the load on the engine is lessened, and the tendency of the engine is to quicken its speed, the expansion-valve acting in unison also quickens its speed in the same ratio as the main-piston; but the main steam-valve continuing to move at the speed regulated by the cataract, the expansion-valve will overrun it, and cut off the steam. On the other hand, if the load on the engine is increased and the tendency of the main-piston is to run slower, the expansion steam-valve will decrease its speed in the same ratio, but the main-valve (continuing to work at the speed regulated by the cataract) will overrun the expansion-valve, and will thus admit more steam as required.

'This motion may be applied to direct-acting engines for various purposes, for which such engines are now applicable. It may also be applied not only to ordinary slide-valves, as

Fig. 36.

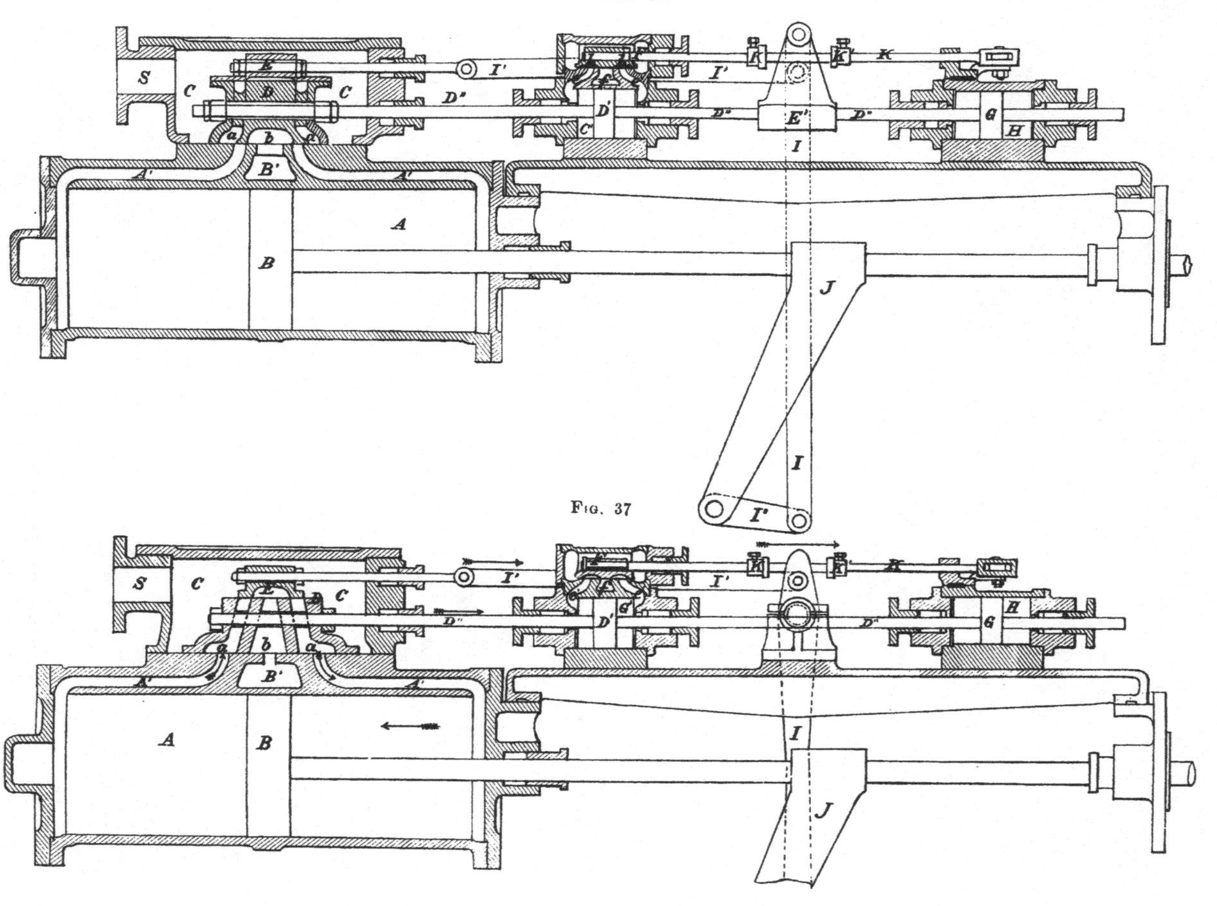

Fig. 37

here more particularly described, but also to rotary and other valves.

'Figs. . . . . [36 and 37] show one modification of my improvements in valve-gear, wherein the expansion or regulating valve, E, receives its motion from the main-piston by means of levers, as represented, and the main-valve, D, receives its motion by the direct pressure of steam in the valve-moving or auxiliary-cylinder, C^1. The motion of the main-valve is regulated by the ordinary cataract, H, through its piston, G, and rod, D^{11}. A is the main steam-cylinder; B, the main-piston; and C, the steam-chest; D is the main-valve moved, as before stated, by the direct pressure of steam upon the valve-moving piston; D^1 is the auxiliary-cylinder; E is the expansion or governing valve worked on the top of the main-valve, D, and moved by motion taken from the main piston-rod through the arm, J, link I^{11}, lever I, and rod I^1, as before mentioned; F is the auxiliary-valve of the B form, and provided with double cavities, *h*, *h*, for admitting and exhausting steam from the auxiliary or valve-moving cylinder, C^1, for moving the main-valve, D, in the right direction; it is operated by the contact of the lever, I, with the tappets, *k*, *k*, upon the rod, K.

'In operating the valves the lever first moves the auxiliary valve, F, admitting steam to the valve-moving cylinder, C^1. The valve, D, is then moved until it establishes communication between the steam and exhaust ports, A^1, B^1, through the cavity, *b*, and brings one of the ports, *a*, into communication with the other main steam-port, A^1, and with the cavity under the valve, D. High steam then passes through the ports, *a* and A^1, in the valve, D, to one end of the cylinder, and is exhausted from the other end through ports, A^1, B^1, and cavity, *b*. The main-piston then commences its motion, and if it move at the same relative velocity as the valve-moving piston, the relative position of the valves will be unchanged, and the admission of steam will be uniform; but, if the speed of the main-piston becomes greater or less than that of the valve-moving piston, the lever will at once begin to act on the valve, E, cutting off or admitting a larger supply of steam, as may be required, the cut-off valve, E, producing the governing or regulating motion, as hereinbefore described.'*

COLEBROOK'S PATENT STEAM-PUMP.

This pump, the invention of Mr. Cemer Thomas Colebrook, of London, and patented March 25, 1874, consists of a steam-cylinder surmounted by a cylindrical piston valve-chest, in the side of which and midway between the two ends

FIG. 38.

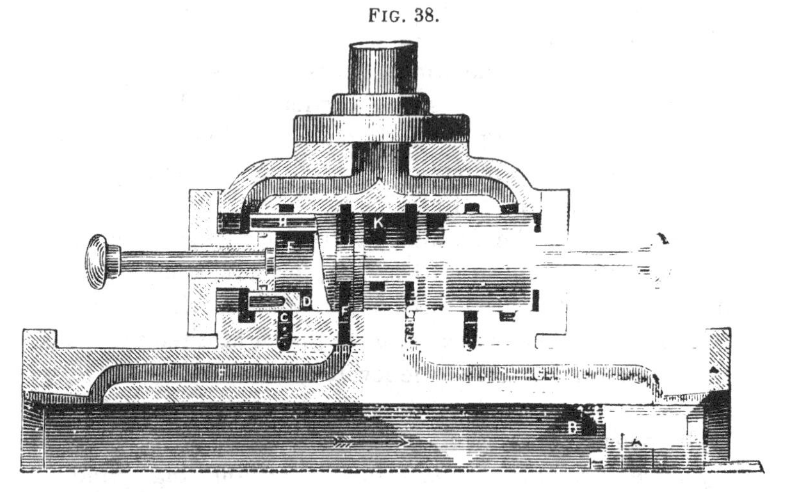

is the main steam-inlet, K (fig. 38). In each end of the valve-chest is a fixed piston, which forms part of the stopper, the diameter of which is about two-thirds that of the valve-chest. Through the centre of each piston is a small spindle, which acts as a starting-handle. Above the valve-chamber, near each end, is an exhaust-port, which terminates in a main exhaust-branch. The slide-valve is formed of two pistons connected by a distance-piece. Each end of these pistons is made hollow to work over the fixed piston at each end of the valve-chamber. The difference in the diameter of the fixed pistons and the valve-chest, as seen

* 'Specification filed by Robert Luke Howard, of Tottenham, in the County of Middlesex, in the Great Seal Patent Office, on the 13th March, 1877.'—*Improvements in Valve-Gear for Direct-Acting Steam-Engines.*

in the illustration, is sufficiently great to permit of the annular projection of the piston-heads being cast hollow. At the bottom part of each projection is an aperture, which communicates between the auxiliary passages and the space between the fixed pistons and the solid ends of the valve. The main steam-piston is of the ordinary type. The main-cylinder is furnished with two main steam-ports, F, G, and two auxiliary or puff-ports, B, L (the illustration should show an auxiliary-port, L, at the other end of the cylinder), the object of the latter being to conduct a portion of the steam which has already done its duty from the steam-cylinder to the valve-chambers, E M, to start the slide-valve, the motion being continued and completed by the exhaust-steam from the main cylinder through the annular ends of the slide-valve, on its way to the exhaust-pipe.

'The positive system is approximately the character of the valve motion of the Colebrook pump before us, in which the action of the high-pressure steam is quickly followed, and aided by the creation of an exhaust at the other end of the valve-chamber. Referring to the sectional illustration of the valve-chamber—the piston, A, has just completed its travel in the direction of the arrow, and uncovered the port B. This port, as shown by the dotted passages, communicates with C, which was in juxtaposition with the port D, when the valve was over at the other extremity of its stroke.

'The live steam thus passes up through the ports, B, C, and D, into the interior chamber, E, of the valve. The steam-pressure thus set up between the valve and the fixed cover—which enters like a piston, and is ring-packed inside this interior chamber of the valve—suffices to move over the valve to some extent, and by this preliminary motion to bring the port of the chamber at the other end of the valve in juxtaposition with a port corresponding to B, which has by that time become an exhaust-port.

'The onward travel of the valve is thus accelerated and determined, whilst by the same motion the port C, as shown in the illustration, is now closed against the disturbing effect of the exhaust immediately set up on the former steam side of A. The corresponding ports to D and C are now in juxtaposition at the other end of the valve-chamber, ready for a repetition of the reversing action when the piston A shall have arrived at the other end of the cylinder. The continued action of the valve is thus maintained by a suitable juxtaposition of ports without shock or jar. The valve itself is built after the principle of an ordinary D valve, sliding over the usual steam-ports, one at either end of the cylinder, but with a steam-passage in the centre, and exhaust exits at each end. As the valve stands in the illustration, the port F to the cylinder has become the exhaust, and G the steam-port to reverse the action of the piston. The exhaust steam issuing through F, passes through the space H, and thus out into the exhaust-branch and pipe at the top of the valve-chest.

FIG. 39.

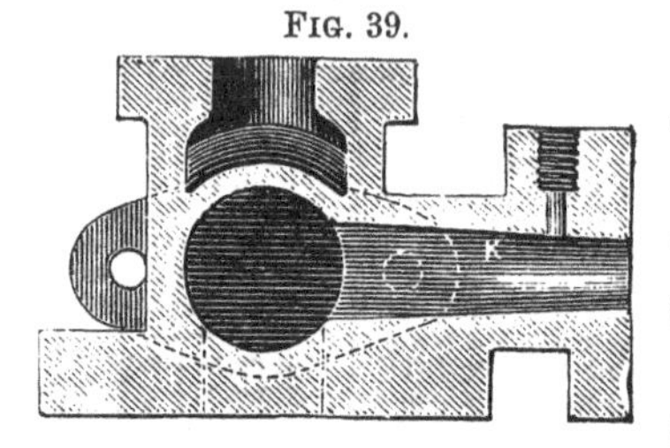

'The steam entry is by the pipe K, shown in the cross section of the valve-chest (fig. 39), and which admits the steam into the annular recess K, in the centre of the valve. From thence, as the valve stands, the steam passes down the port G to the cylinder to reverse the action of the piston.

'The hand-tappets are provided to enable the attendant to move over the valve at starting should it have stopped on a centre with all parts closed or open. The action of this valve is very simple, and should be satisfactory, as there is only one moving piece, and the reversal by a suitable cushioning adjustment ought to be elastic and noiseless.' *

Fig. 40 is an illustration in perspective of one of these pumps for feeding boilers or for light pressures. Fig. 41 is an elevation, fig. 42 a half sectional plan, and fig. 43 an enlarged section through the valve-box, showing the arrangement of the gun-metal valves. These valves are made of such a size that, by merely unbolting the one top cover, both

* *Iron*, August 30, 1879.

valves can be taken out and examined, the delivery or top valve being made so much larger than the suction or bottom

FIG. 40.

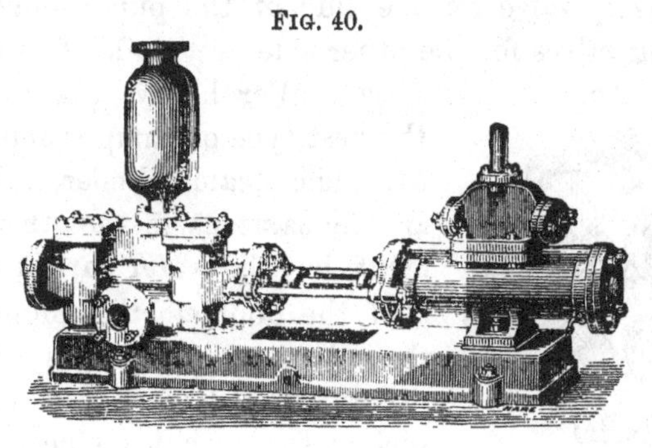

valve that the latter can be taken out through the delivery-valve seat.

FIG. 41.

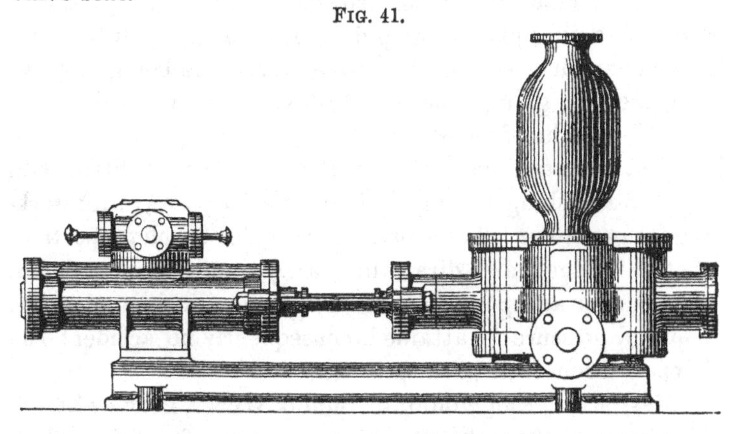

For pumping cold water moderate lifts, the valves are made of leather, india-rubber, or canvas, according

FIG. 42.

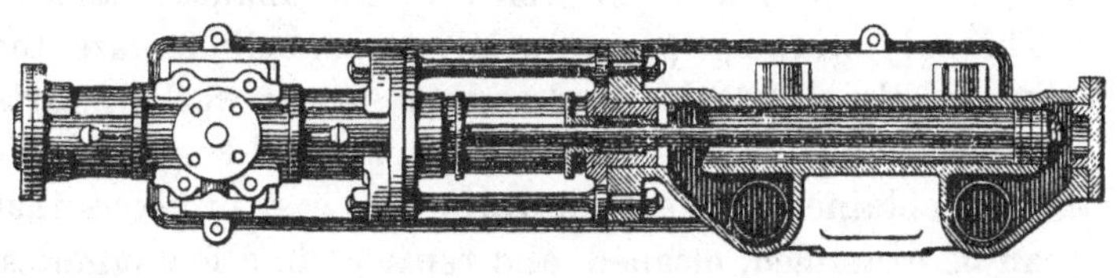

to the nature of the fluid to be pumped. They are made in two ways; in one arrangement the two delivery-valves are made of one piece, and the two suction-valves are also made of one piece. In the other arrangement one suction and one delivery-valve for one side of the pump and the corresponding valves for the other side are made of two separate pieces. For heavy lifts in collieries the best type of pump is shown in fig 44. The steam-cylinder and its gear are the same as in the other arrangement, but the water part of the pump is of the double-acting plunger type; the valves are of the class called wing-valves, furnished with india-rubber springs to prevent undue concussion, and thereby increase the durability of the valves.

FIG. 43.

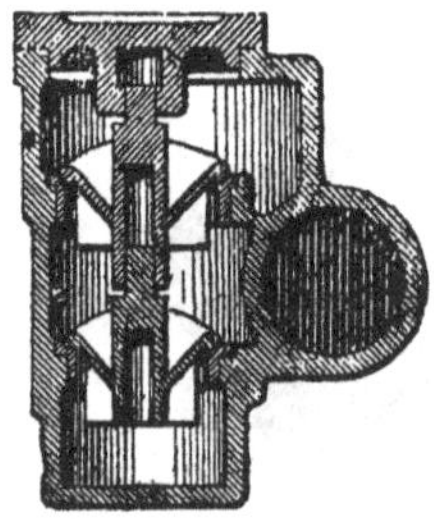

Fig. 45 represents a Colebrook's Patent Pump fitted with a double-acting piston-pump designed for heavy lifts; there are no flat surfaces but the covers, all others being circular, and thereby giving the greatest strength with the least possible weight of metal.

Fig. 46 is an illustration of the same patent, fitted with a piston and plunger-pump for hydraulic purposes, or for working accumulators. These pumps are made of such proportions that in forcing water direct into a hydraulic-press cylinder, the cylinder and pump-pistons are in equilibrium when the required pressure is attained, consequently no accident can happen through its great pressure.

Messrs. May and Mountain, Suffolk Works, Berkley Street, Birmingham, the makers of this pump, claim for it the following advantages: No tappets, valves, eccentrics, levers, or other mechanical appliances are used to actuate the steam slide-valve, an office which is performed by the exhaust steam.

The only working parts in the steam-cylinder are the piston and the slide-valve, and as there are no working parts in either the piston or cylinder-covers, the full length of the stroke is obtained. The slide-valve is so easy of access that it can be examined, cleaned, and replaced in a few minutes, and as it is immaterial which way it is inserted in the valve-box, no error can be made in replacing it. The piston is

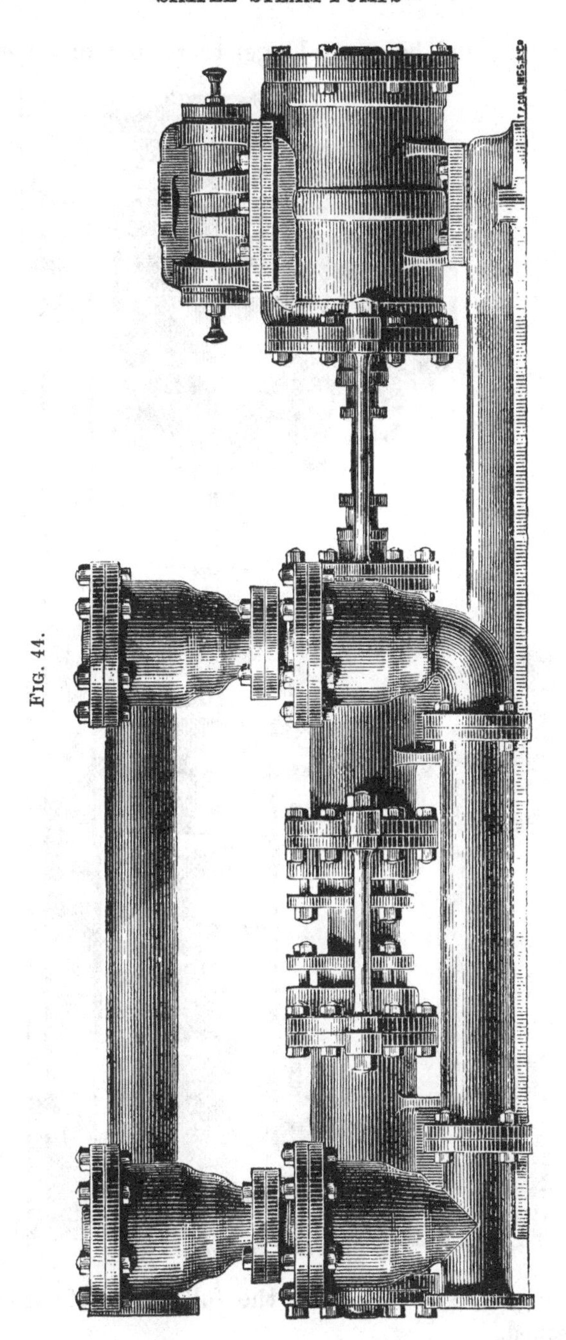

FIG. 44.

quite plain, and being no longer than that of an ordinary

FIG. 45.

steam-engine, a stroke nearly the full length of the cylinder is obtained.

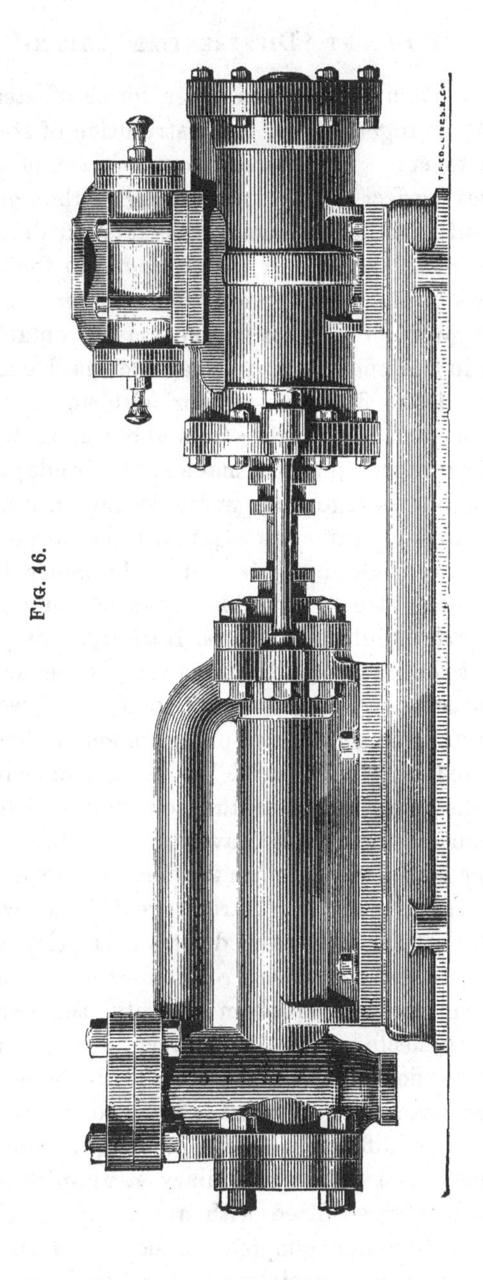

FIG. 46.

DAVEY'S PATENT 'DIFFERENTIAL' STEAM-PUMP.

This invention differs from other forms of steam-pumps in the valve arrangement for the distribution of steam, which is such as to suit every increase or decrease of resistance, and secures perfect safety in working, although the load may suddenly and greatly vary. A detailed description of the 'Differential' valve-gear will be found in the account of Mr. Davey's Compound Engines. The larger sizes are provided with governors. This engine may be obtained with a condenser in the suction-pipe at an increased cost, varying from about 13 to 31 per cent., or with an air-pump condenser at an extra cost varying from about 22 to 40 per cent. Another form of this pump is made specially adapted to irrigation, pumping sewage, and general surface drainage, where large quantities of water are required to be raised a limited height, and are made in all sizes up to 36 inches diameter of pump and 5-feet stroke. Fig. 47 shows the pump.

'Among the exhibits of Messrs. Hathorn, Davey, and Co., of Leeds, at the Paris Exhibition (1878), is the new form of Differential Steam-Pump (Davey's Patent) which we illustrate above. In their former steam-pumps of small sizes this firm had only used the "differential" valve-gear in an incomplete form for cheapness' sake; in the new type which is represented in our engraving they have succeeded in applying the complete gear in a way at once simple and inexpensive.

'The principle of Mr. Davey's gear is, as will be remembered, that the slide-valve derives its motion both from the main crosshead and from a rod connected with a cataract-piston moving with a uniform velocity, the two motions being so connected that the cut-off depends upon the velocity of motion of the main crosshead—in other words, upon the resistance encountered or work to be done by the piston. In the differential pumping engine the cataract-piston is connected with a subsidiary steam-piston, working in a separate cylinder fitted with a small slide-valve. The principal simplification adopted in the new form of pump is the making of this subsidiary piston (two being now used)

Fig. 47.

in one with the main slide-valve, its cylinder being a part of the slide-valve chest, on the side of which the auxiliary slide-valve works.

'The whole arrangement is shown clearly in our engraving. The slide-valve (a common D) is placed above the cylinder, and in one piece with it are two pistons (one at each end) working in cylindrical prolongations of the valve-box. The outer end of these cylindrical chambers are connected by ports with a small valve-chest placed beside the other, in which works a small D-valve. The main-valve is connected by a link to the upper end of the working lever (its stroke being adjustable by varying the position of the connecting-pin in the slot shown); a pin a few inches lower down the lever is fixed to the crosshead of a cataract-piston, and guided in a straight line, while the lower end of the lever is connected to an arm on the main-piston and pump-rod. The subsidiary-valve is worked from the lever by the button and stop arrangement shown. The action of the gear is easily understood: the auxiliary-valve, being opened by the stop at the end of the stroke, admits steam behind one of the pistons and connects the other with the exhaust; this steam opens the main-valve, and keeps it moving forward at a rate determined by the cataract-piston, which, of course, commences to move at the same time as the main-slide. Steam is thus admitted to the main-piston, and the lower end of the working lever commences to move with it, the tendency of its motion being to *close* the valve. Cut-off then takes place, as in the differential pumping engine, at a point determined by the rate at which the closing motion of the crosshead overtakes the opening motion given by the steam acting directly on the main-slide.

'The automatic control which this gear exercises over the motion of the pump is as perfect as in the differential pumping engine, even extending, like it, to the case of total removal of the load from one side of the piston by any accident. The full speed of the pump is 120 feet per minute, but it can work, as we have ourselves seen, quite steadily at two or three strokes per minute if required.

'The pump has a diameter of 8 inches and a stroke of 2 feet, the diameter of the steam-cylinder being 12 inches. The maximum delivery is 250 gallons per minute. The pump-valves are india-rubber discs, which the makers use for heads up to 150 feet, above that employing single or double-beat valves with leather beats. The workmanship of the pump is thoroughly good, and we understand that it is made at a price but little above that of the ordinary direct-acting pumps, without any controlling or expansive arrangements.'*

THE 'BLAKE' PATENT STEAM-PUMP.

This pump is the invention of Mr. G. F. Blake, of Boston, and was produced in America about eighteen years ago. We have been favoured by Messrs. S. Owens and Co., of Whitefriars Street, Fleet Street, London, E.C., with a drawing of the first Blake pump, bearing the date 1862. The manufacture of this pump was begun in this country about the early part of 1875, solely by the above-named firm. Up to the year 1877 nearly 11,000 of these pumps had been made in America. The accompanying illustrations represent two forms of this pump: fig. 48 is a piston-pump suitable for moderate lifts, fig. 49 is a double-plunger pump adapted to high lifts and water containing solid impurities.

Referring to the sectional elevation, fig. 50, it will be observed that the main steam-cylinder is surmounted by a small secondary-cylinder, which contains an auxiliary-piston—a double-headed spring-ring piston—for actuating the main D slide-valve, which fits between two collars, by which it is mechanically moved. The main-valve works on the top of a three-ported auxiliary slide-valve contained in a chamber in the bottom of the auxiliary-cylinder.

It will be observed that the ports of the auxiliary-valve correspond with those of the main-cylinder. Into each end of the auxiliary steam-cylinder is fixed an immovable horizontal arm, to which is anchored a vertical lever. Fixed to the auxiliary-valve are two long spills or rods, the outer

* *Engineering.*

ends of which take hold of the aforesaid levers from which the valve obtains its motion. The motion which the levers receive is imparted by tappets in the main-cylinder ends,

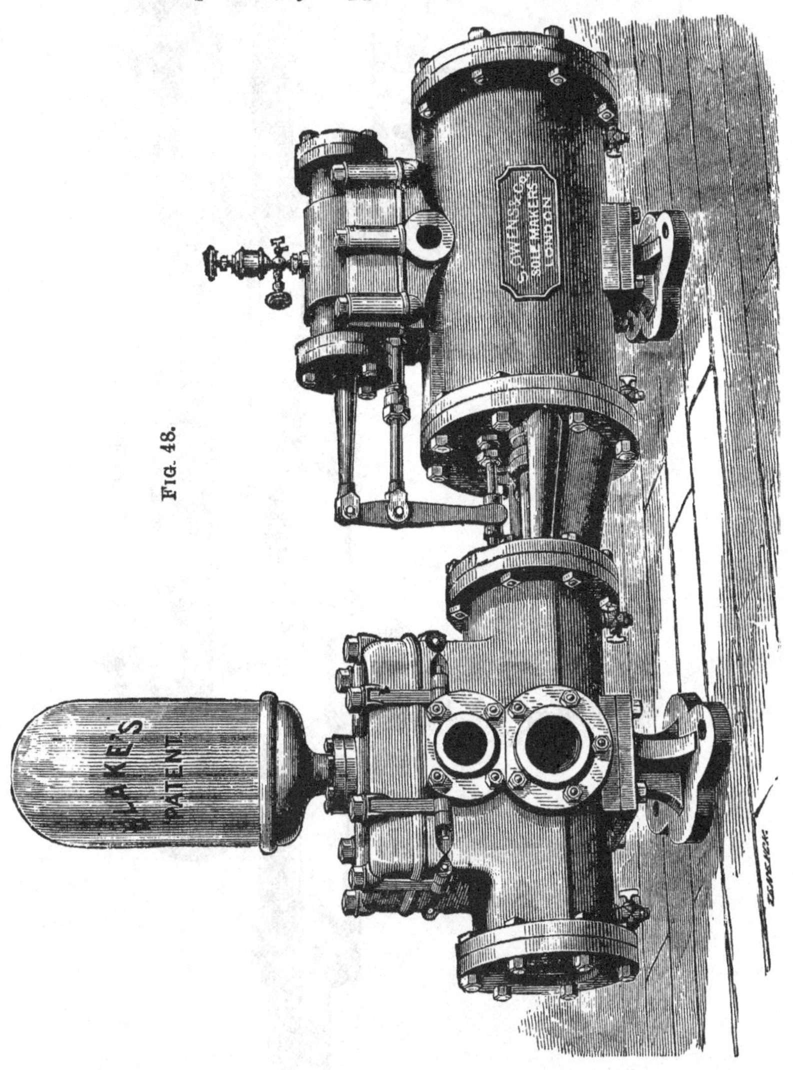

Fig. 48.

which are moved by the piston. In some forms of this pump, to make it more compact, the tappet in the back end of the main-cylinder is dispensed with, the motion which it gave

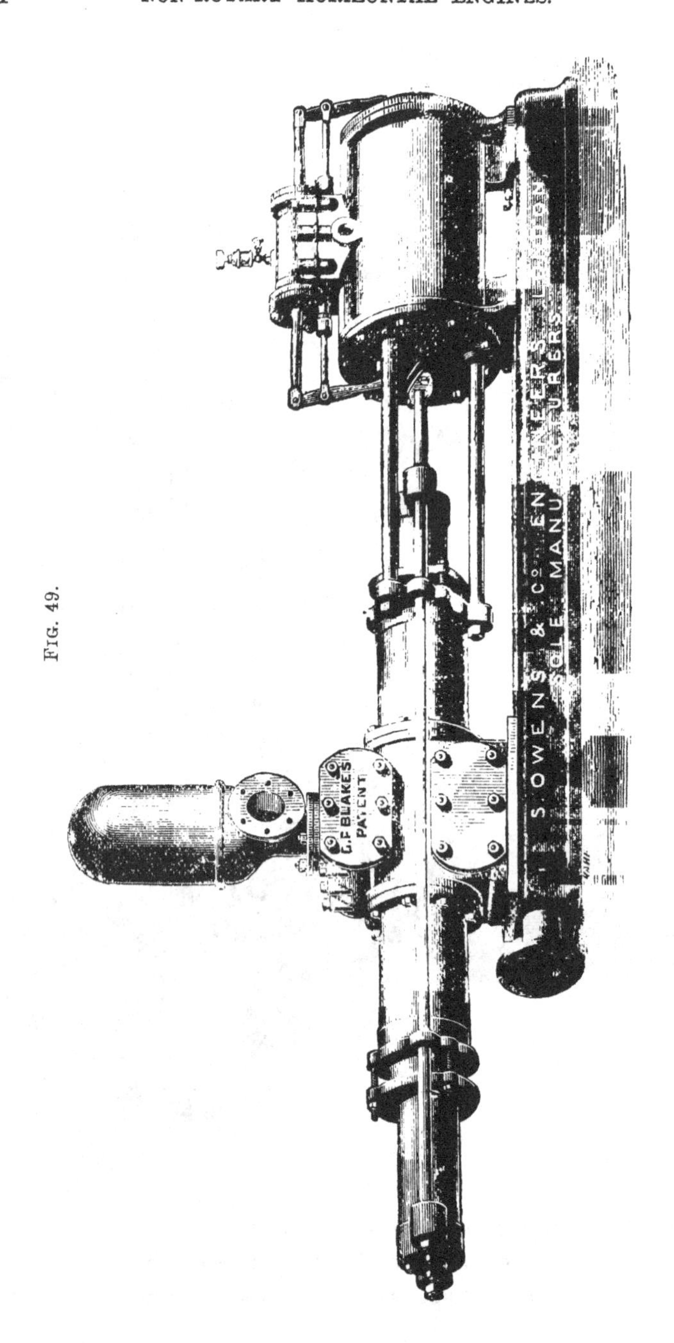

FIG. 49.

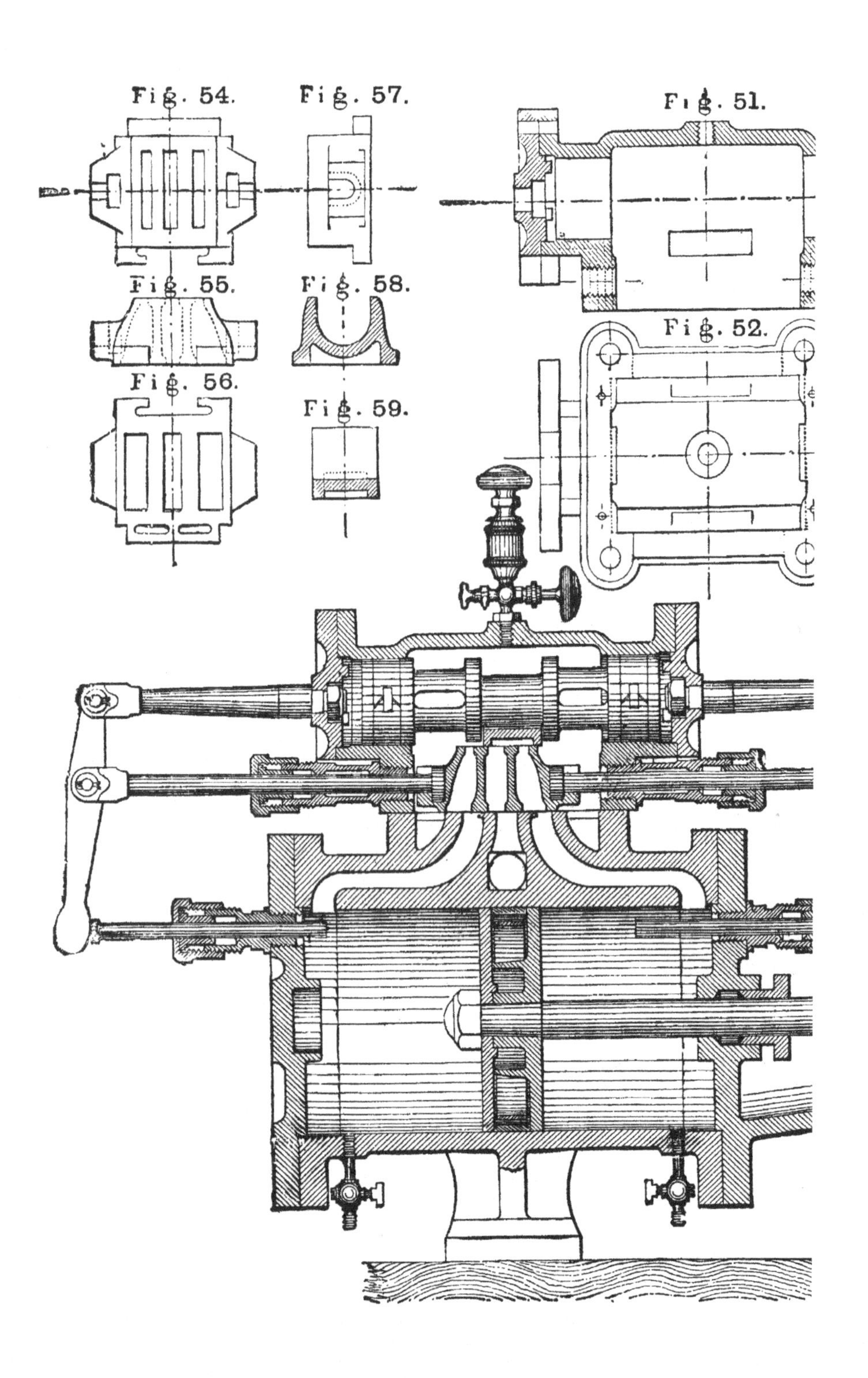
Fig. 54.
Fig. 57.
Fig. 51.
Fig. 55.
Fig. 58.
Fig. 52.
Fig. 56.
Fig. 59.

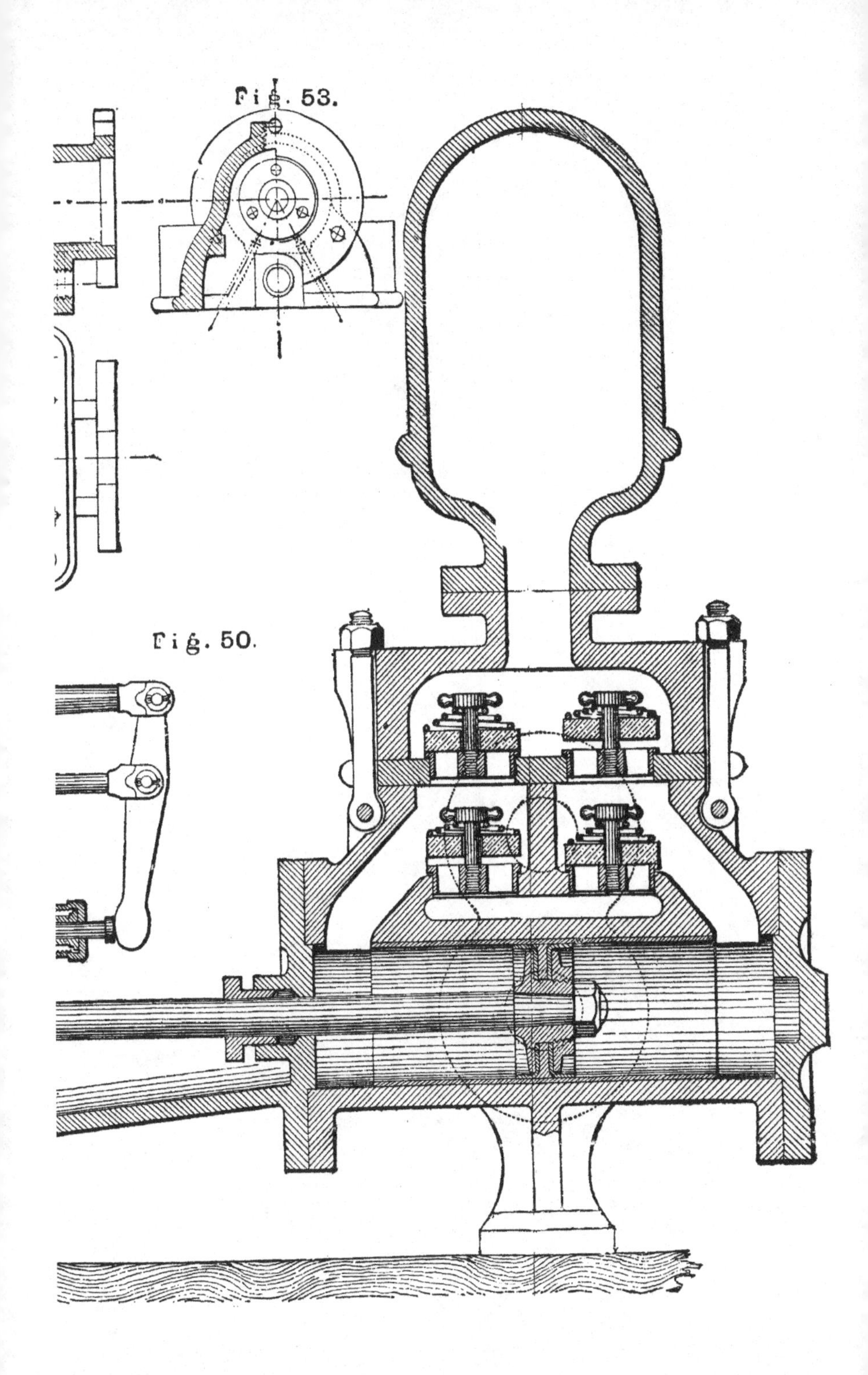
Fig. 53.
Fig. 50.

being contributed by a tappet in the water-cylinder, wherein it is actuated by the pump-piston. This will be observed in the diagram which accompanies the explanation of the action of the pump.

The following detailed description of this pump is taken from 'Engineering' of July 9, 1875.

'Our engraving shows at fig. 50 a longitudinal section of the Blake steam-pump; fig. 51 being a section of a secondary cylinder, which contains an auxiliary-piston for actuating the main-valve; whilst figs. 52 and 53 are respectively a plan and half-end elevation and cross-section of this cylinder. Figs. 54 to 57 show a plan, elevation, underside view, and end elevation of the auxiliary-valve; figs. 58 and 59 being sections of the main-valve. Both the main and auxiliary-valves are plain flat slides, the main-valve being a common D-valve, and the auxiliary a valve of the form shown. The latter, being attached to a rod which receives an impulse from the main steam-piston, is moved with the same absolute certainty as is the slide-valve of an ordinary engine driven by an eccentric.

'It will be seen that the secondary-cylinder is mounted on the primary or main-cylinder. The ordinary spring-ring steam-piston which it contains drives the main slide-valve which works on the upper face of the auxiliary-valve. This valve has three ports of equal area, which correspond in every position with the ports of the main-cylinder.'

'In working, if the main-piston should attain a velocity in excess of the piston which actuates the main-valve, the piston strikes the tappet, which is seen in fig. 50 projecting through the cover into the main-cylinder. By this means a lead is given to the main-valve, steam being thereby admitted in front of the piston, forming a cushion, and giving steam to start the piston on its return stroke. It will be observed that the auxiliary-valve has two slots cut in its underside as in fig. 56. These slots communicate with the main exhaust-passage, and also give steam from the valve-chest to both ends of the auxiliary or main-valve piston alternately. The result of this is, that directly the auxiliary-valve is thrown over by the action of the tappet-rod, steam is given on one side of the auxiliary-piston, and exhaust takes place on the other. On the opposite side of the auxiliary-valve—the upper side in fig. 56—another slot is formed, which at the right moment enables a small quantity of steam to pass to the exhaust side of the auxiliary-piston, and so to form a cushion to prevent it striking the cylinder-cover. In these combined operations no waste of steam occurs, as it is retained and gives out its useful effect on the return stroke. The result of this ingenious combination of valves is a perfectly continuous action without any dead point and unassisted by extraneous means. This result is attained without any complex internal arrangement, and without the presence of any parts which are liable to get out of order. The valve-gear is, in fact, most simple in its character, as a glance at the engraving will show. Of the excellent working of the Blake pump at extreme ranges of speeds, we are enabled to speak from personal observation, having inspected its working at Messrs. Owens' works. The apparatus was a 5-inch pump, with an 8-inch steam-cylinder, having a 12-inch stroke. It was started to work at 155 single strokes per minute, and the speed was varied down to 25 strokes. It was then again run up to a high speed, and suddenly set to a speed which for slowness and perfect continuity has never been equalled in our experience. With the delivery throttled to represent a head-pressure of water of 230 feet, and with an average steam-pressure of 40 pounds, this pump ran, or rather crawled, at the rate of one stroke in 12 minutes, or 5 strokes per hour. delivering water throughout. We thus have a piston speed of 1 inch per minute, or 5 feet per hour, and a continuous delivery of water.

'It is needless to observe that such a slow speed as this could not possibly be required in practice, but it illustrates the reliability of the pump either in quick or slow working. It is also needless to observe that we did not wait to see many of these slow strokes, as by the time the engine had performed two, occupying 24 minutes, we were satisfied. The pump was then put to work under rapidly varying loads, and throughout acquitted itself most creditably, affording ample foundation for the opinion we have already expressed that it will not fail to establish for itself as high a reputation in England as it has in America, where we understand that it is the only direct-acting steam-pump used in the navy.'

The operation of the valve-gear and the general action of the pump will be understood by the following description

FIG. 60.

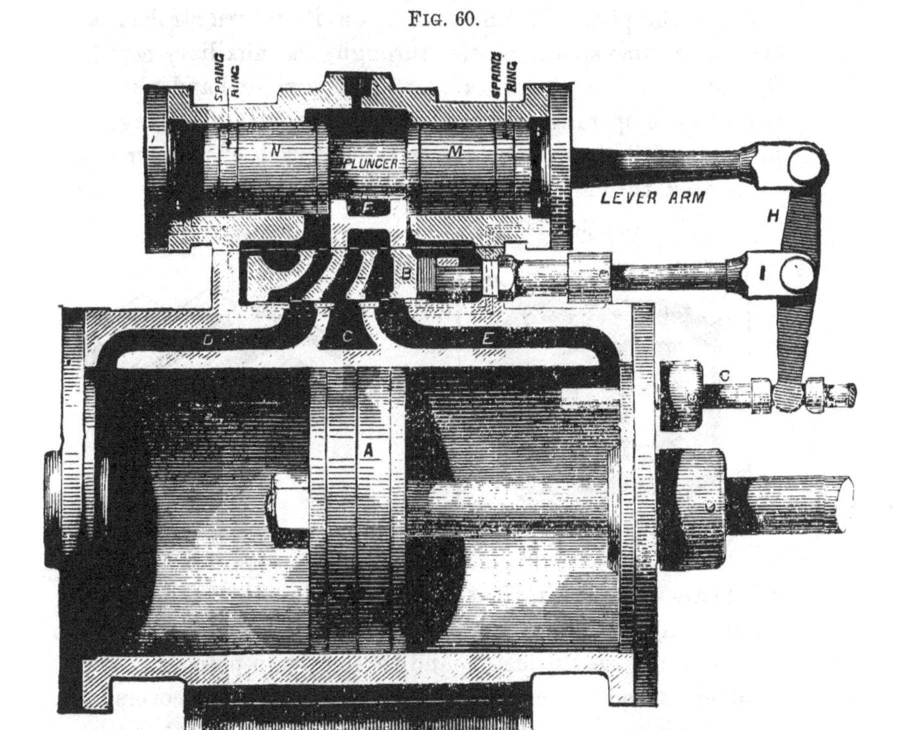

given by the makers, and the sectional drawing, fig. 60, of the steam-cylinder and the valve-box, and figs. 61 and 62:

'In order to understand the action of the valve, suppose the piston (A, fig. 60) to be moving to the right, the movable seat, or auxiliary-valve (B), will then be at the extreme left, with the exhaust-port (C) open on the right-hand stroke, as shown in the illustration, and steam on to the left-hand through the port (D); directly the piston (A) approaches the end of its stroke it operates the tappet (G), which communicates motion through the lever (H) and the rod (I) to the movable seat and auxiliary-valve (B). By this operation steam is at once given to the right-hand side of the piston (A) through the port E, which it slightly opens in sufficient quantity to cushion the piston (A) and start it on its return stroke; at the same time steam passes through the auxiliary-port (J, fig. 61), which communicates with the right-hand plunger (M), at once opening the main slide-valve (F) and giving the piston (A) full steam; the steam at the back of plunger N is exhausted through L and K to the main exhaust (C). The

Fig. 61. Fig. 62.

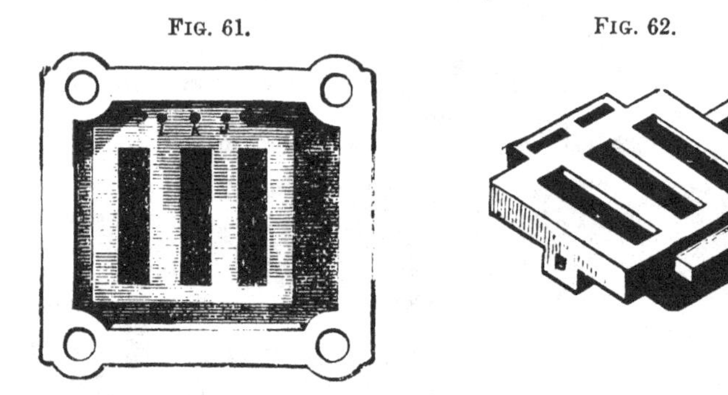

valves are operated in the same way at the other end of the stroke, only in the opposite direction, thus giving continuous action to the steam-piston and to the pump-plunger. The plungers M and N are prevented from striking the covers by an ingenious arrangement of ports and valves at either end, which checks them with the utmost certainty.'

The following remarks by the makers relative to the most noteworthy features of the 'Blake' pump, the advantages they claim for it, and the description of some of the principal parts, are inserted as useful addenda to what has already been stated of this excellent and well-known machine:

'Its simplicity of construction, positive action under any pressure, the perfect arrangement of all its parts, enable it to run at any speed with certainty, delivering more water, with a less expenditure of steam, than any other direct-acting steam-pump.

'An important feature, peculiar to the "Blake" pump, is the mechanical connection between the movable seat (which in certain cases becomes the main-valve) and the main-piston. It permits the pump to be run at a very high speed without danger of striking the heads. The piston cannot reach the end of the cylinder, as the mechanical connection referred to gives a lead as certain as can be obtained with an eccentric.

'It will start at any part of the stroke by simply admitting steam into the cylinder of a pressure in excess of the resistance on the pump-plunger; will work at any speed both with high and low-pressure steam; has no starting or operating-handle to assist its action (such being unnecessary in this pump, as there is no dead point at any part of the stroke); and it cannot be stopped, unless by shutting off the steam, so long as the pressure exceeds in the steam-cylinder the load or resistance on the pump.

'The novelty in the "Blake" steam-pump is its perfect and ingenious arrangement of steam-valves, which meet all the conditions of a direct-acting steam-pump, viz., simplicity, durability, positive action, and the highest speed attainable. It is a combination of two slide-valves, working as perfectly when the exhaust is connected with a condenser as when passing steam into the atmosphere.

'The main-valve is placed in such a position as to be driven by a spring-ring steam-piston, that becomes necessarily as positive in action as any steam-piston when exposed to pressure. The auxiliary-valve is a plain flat slide-valve attached to a valve-rod, which receives its motion from the main steam-piston; it is therefore operated with the same degree of certainty as an eccentric moves the slide-valve of an ordinary steam-engine. The casting forming this auxiliary-valve has three ports, which coincide in every position with the three ports of the main-engine, thus forming an upward extension of the engine-ports, on the upper seat of which the main-valve slides. If the main-piston should attain a velocity exceeding that of the piston which actuates the valve, it would strike the cylinder head; but the movable valve-seat makes such a contingency an impossibility, because having a mechanical connection to the valve-rod, it is brought into such a position as to become the main-valve, independent of the action of the main-valve proper, and gives direct steam to cushion and reverse the engine.

'The presence of a small cylinder surmounting the main-cylinder must not be confounded with that adopted by many other manufacturers. This cylinder is simply for the purpose of containing an ordinary spring-ring steam-piston, not a valve. In many other direct-acting pumps it contains an auxiliary or secondary-valve.

'The main and auxiliary-valves being merely flat plane surfaces, and their respective positions being face to face, the wear is consequently even as well as compensating, the same as any plain steam-engine valve. As a protection of the valve-gear from wear, at each joint may be found a steel friction-roller that, in the event of its wearing out, can be renewed at a trifling expense. All the parts of the valve-gear being simple in form and made of wrought-iron, in case of accidental breakage any of them can be made in a blacksmith's shop without the aid of either machine or skilled labour.

'The pistons are fitted with adjustable spring-rings, which compensate for wear. The steam-piston is so arranged as to permit the packing-rings to revolve in the cylinder; the wear therefore is necessarily more even than in the older plan of stationary ring-packing. All the piston-rods are made either of steel, gun-metal, or copper.

'All stuffing-boxes are made of the best gun-metal; in the event of one, or more, receiving injury by either breakage or wear, it may be detached from the pump and a duplicate part replaced in a few moments.

'Every part of the machine is made to uniform gauges, and any part worn out or otherwise injured can be promptly supplied.'

THE 'CALEDONIA' PATENT STEAM-PUMP.

This pump was introduced in 1874. The steam-valve cylinder, as in other pumps of this class, is mounted upon the

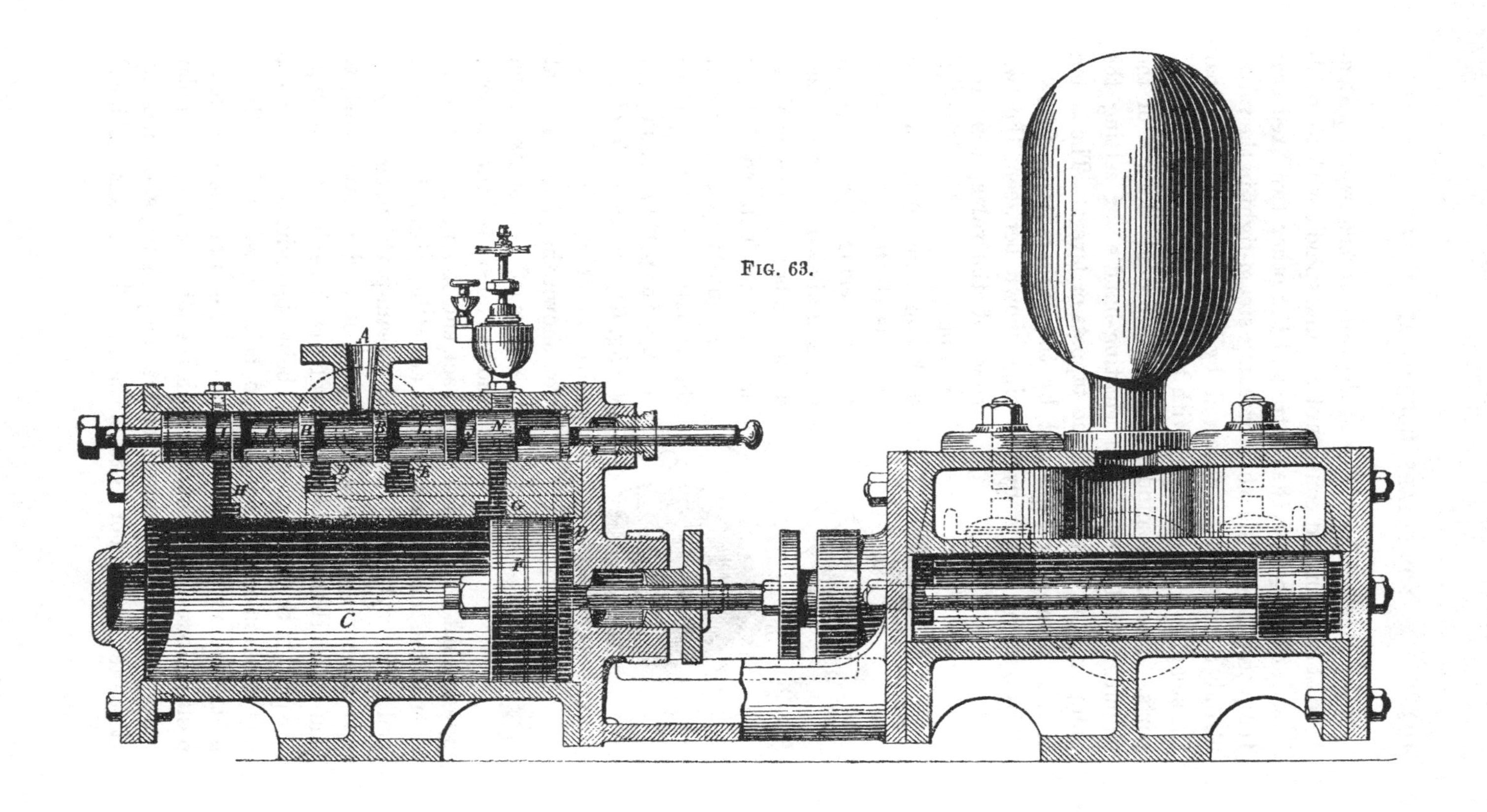

FIG. 63.

main-cylinder. The valve is of the annular form, with a piston-head at each end; it consists of a round spindle with four collars, two being near the centre and the other two fixed near the piston-heads, the whole working steam-tight in the valve-chamber. It will be observed, by reference to fig. 63, that the steam-cylinder is fitted with two steam and two exhaust-ports. The valve-spindle is carried through the end of the chamber, and serves as a starting-handle for setting the pump to work. There are no tappets or levers. The action of the pump, as explained by the makers, is as follows: 'The steam enters by the pipe shown between the two rings of the valve, B B, and is admitted to the cylinder, C, through the ports, D E. The valve is shifted by the piston, F, uncovering the jet-ports, G H, and allowing steam to act on the ends of the valve alternately through the small holes in the annular spaces, I and J, the exhaust all taking place through the ports, K L, and pipe, M (fig. 64), in the side of the valve-chamber. In the position shown the piston, F, is at the forward end of its stroke, and has uncovered the jet-port, G, and shifted the valve to the back end of its stroke, and has opened the main steam-port, D, admitting steam to the front end of the cylinder, C (the main steam-ports being crossed), and exhausting through the port, E, and exhaust-port, L. The steam that has shifted the valve is also exhausted through the port, L, the annular space, J, being brought over the port, and remains open, the jet-port, G, being closed to the cylinder by the hollow part, N, while the other end of the valve is in position to receive steam through the jet-port, H, and annular space, I, when the piston uncovers the jet-port, H, on arriving at the termination of its stroke back. The valve is cushioned

FIG. 64.

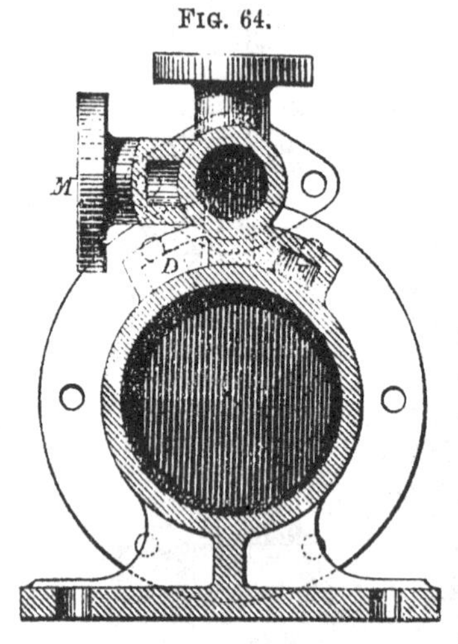

by the exhaust-steam being allowed to get behind the valve through small holes; these holes also serve to exhaust more quickly, and allow the pump to be run at a higher speed. These pumps have been run at 200 feet per minute down to 20 feet with equal certainty, and with no knocking of the valve.'

The makers of this pump, Messrs. A. F. Craig and Co., of Paisley, profess that it occupies less space than any other direct-acting steam-pump, the cylinder being the same length as for any other engine of equal length of stroke.

The 'Selden' Steam-Pump.

This steam-pump was introduced about nine years ago. The annexed illustrations, figs. 65, 66, and 67, represent an elevation, a longitudinal section, and a back elevation. It will be seen that, as in other pumps of the class, the main-cylinder is surmounted by the valve-chamber. The bottom of the valve-chamber contains a main D slide-valve and two starting slide-valves, D D, one on each side of the main-valve, and all working upon the same face. Each starting-valve is attached to a spindle working through a stuffing-box at each end of the chamber. Each spindle is connected to a lever, C, which is actuated by a tappet, B, in the end of the main-cylinder. The top part of the valve-chamber contains an annular auxiliary or shooting-valve, F, with a piston-head at each end, the main slide-valve being attached mid-way between the piston-heads. The main-cylinder contains the usual pair of steam-ports, and the exhaust-branch, covered by the main slide-valve in the centre of its face. At each end of the face is a small steam-port, E (communicating with the ends of the shooting-valve), and an exhaust-branch, H, discharging into the main-exhaust; these are covered by the starting slide-valves. The tappets before referred to receive their motion from the main-piston, which is of the ordinary type. We will suppose the latter to be travelling towards the right-hand end of the cylinder. Near the end of its stroke it strikes a tappet, B, in the end of the cylinder, which gives motion through a

FIG. 65.

double-ended lever, C (anchored to a fulcrum on the cylinder end), to the starting slide-valves, D D, causing them to make

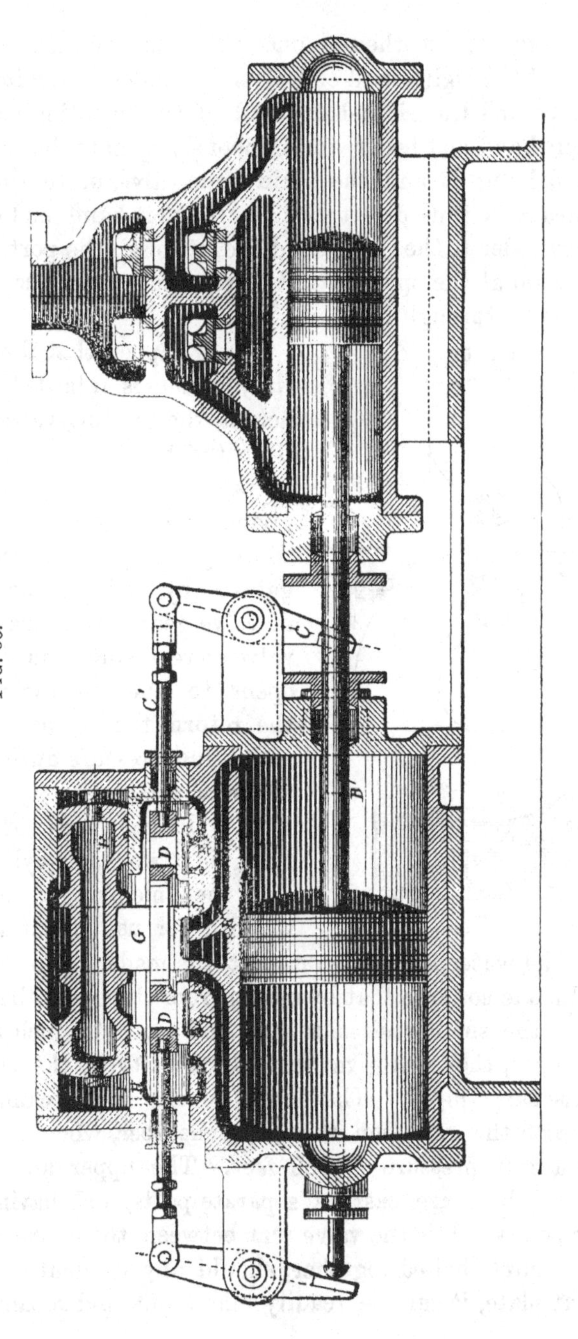

FIG. 66.

a stroke to the left, thereby uncovering the auxiliary steam-port, E, at the right-hand end of the cylinder, and admitting steam behind the right-hand end of the shooting-valve, F (the left-hand end being simultaneously open to the exhaust, H H′), driving it and the main slide-valve, G, to the left. The steam is thus diverted from the left-hand end of the main-cylinder to the right-hand end, through the port, I; the used steam at the opposite end at the same time escaping to the exhaust through the ports, A K.

FIG. 67.

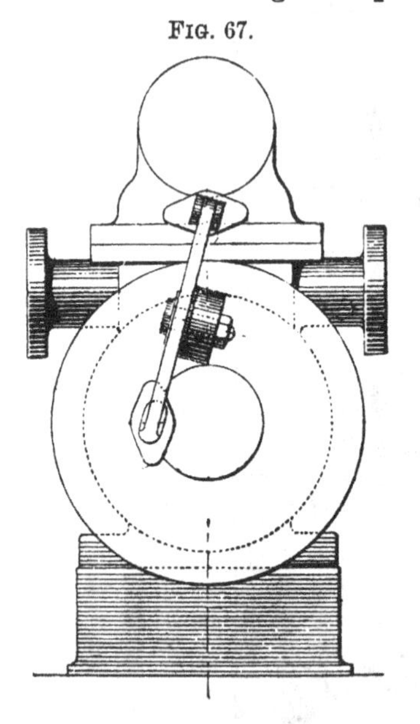

If it happen that the auxiliary piston is retarded in its action, the starting-valve compensates for the irregularity by tapping the main-valve and giving the required cushion to the piston. The advantage gained by this arrangement of valve-gear is that the main-valve on reversing immediately opens to the full extent for the return, thereby giving the piston full pressure during the whole of the stroke.

A plunger type of this machine (two water-cylinders, with the same plunger working in both for use in situations where the water is muddy or gritty) is made. 'The valves are all made so large that by lifting from three-eighths of an inch in the smaller sizes to one and a quarter inch in the larger sizes, they will give the full capacity of the suction and discharge-pipes. The action of the valves cannot be heard with the ear upon the valve-chamber, when working under a test pressure of 350 feet. The upper and lower valve-chambers are cast in separate parts, and having the plate upon which is the valve-seat between them, the whole being securely bolted together, should any accident occur to the seat-plate, it can be readily taken out and repaired or renewed without the loss of any other part. The same steam-valves are used as in the piston-pump.

'The "Selden" pump has often been run with water instead of steam, and it will run as steadily and reverse as promptly as with steam. When under water it will work well. On one occasion there being no steam for several days owing to the boilers giving out, the mine was flooded, yet as soon as steam was let down the pump started promptly, and was run at a higher rate of speed than it had ever been run at before.' (Maker's remarks.)

The makers of this pump are Messrs. John H. Wilson and Co., Bankhall Foundry, Sandhills, Liverpool.

THE 'IMPERIAL' STEAM-PUMP.

This pump is the invention of Mr. J. E. Rogers, by whom it was patented in September 1873. The illustrations, figs. 68 and 69, show an elevation and a sectional elevation of the valve-gear, and a half-section of the main-cylinder. As in other pumps of this class the main steam-cylinder is surmounted with a cylindrical valve-chamber, as will be seen. The main-cylinder is furnished with a short piston of the ordinary type. The valve-chamber contains two plungers or pistons, united by a distance-piece, which actuate an ordinary flat valve. Two small auxiliary-valves are connected outside to the side of the main-cylinder. They are operated by live steam from the main-cylinder, and admit steam to one or other of the ends of the valve-chamber, thereby giving the plunger-valve an impulse to the right or to the left, as the case may be, the main-valve being carried with it. The simplicity of the action of the valve-gear, a very important advantage to fast-running machines, and one not so common to this class of machines as could be desired, will be apparent from the following description. Suppose the piston to be in the position shown in the accompanying figure, having completed its stroke, and having passed the port D, the steam from behind it enters the port, and lifting the auxiliary-valve, A, passes onwards through E to the end of the plunger, C, placing it in the position shown, whereby the port, K, is opened, and the steam is permitted to pass freely to the right of the piston, and also through the port, F, on to the top of the auxiliary-valve, A, which it replaces in

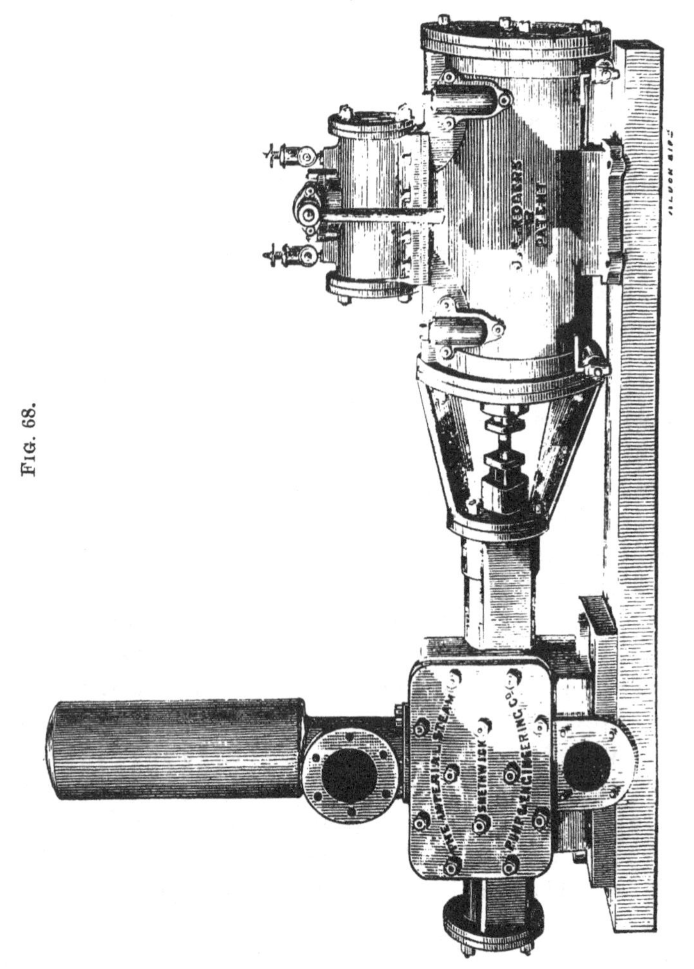

Fig. 68.

its seat, shutting off communication with the valve, C. The same action takes place at the opposite end. An explanation of the details will help the reader to more easily understand the internal arrangement of this pump: A represents the auxiliary-valve; B is a screwed plug, which shuts down on

FIG. 69.

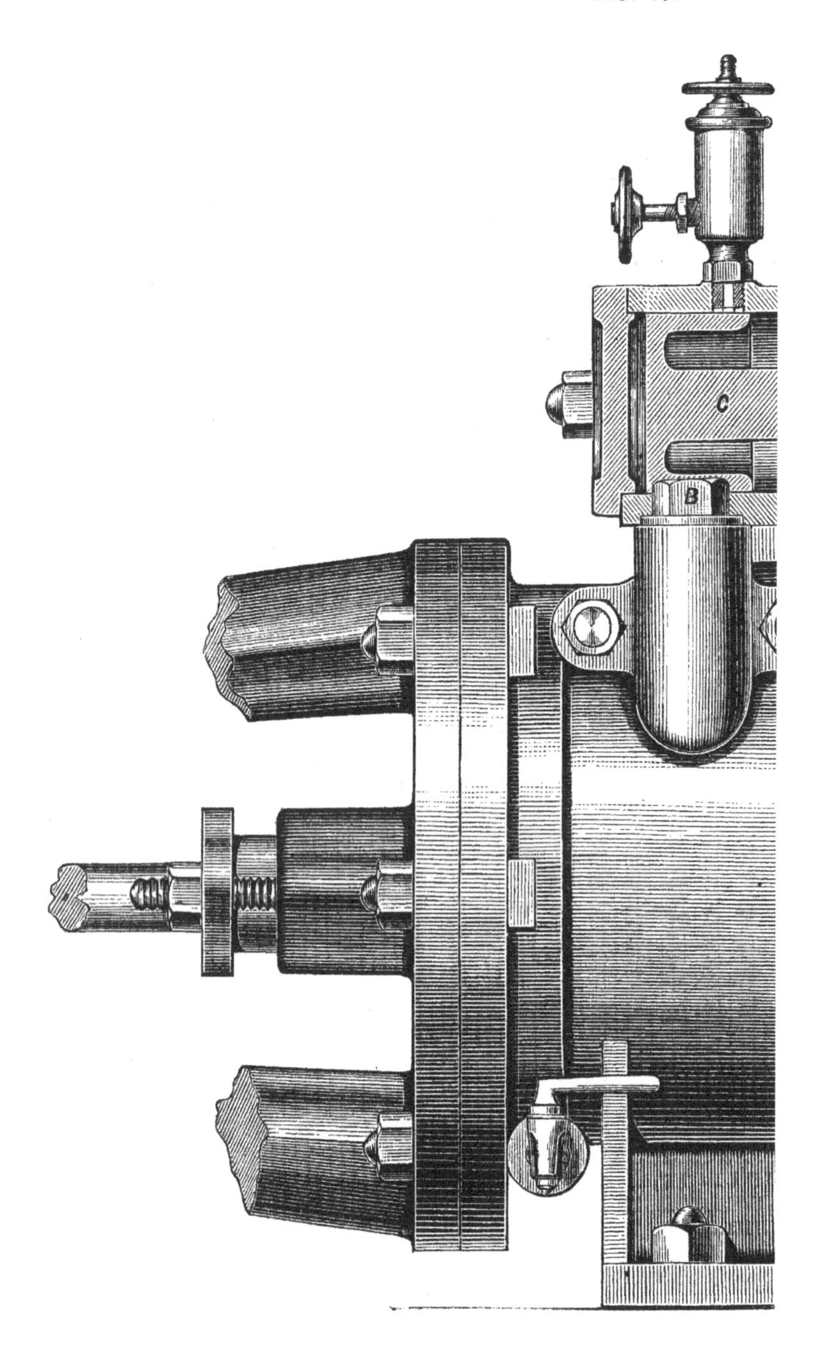

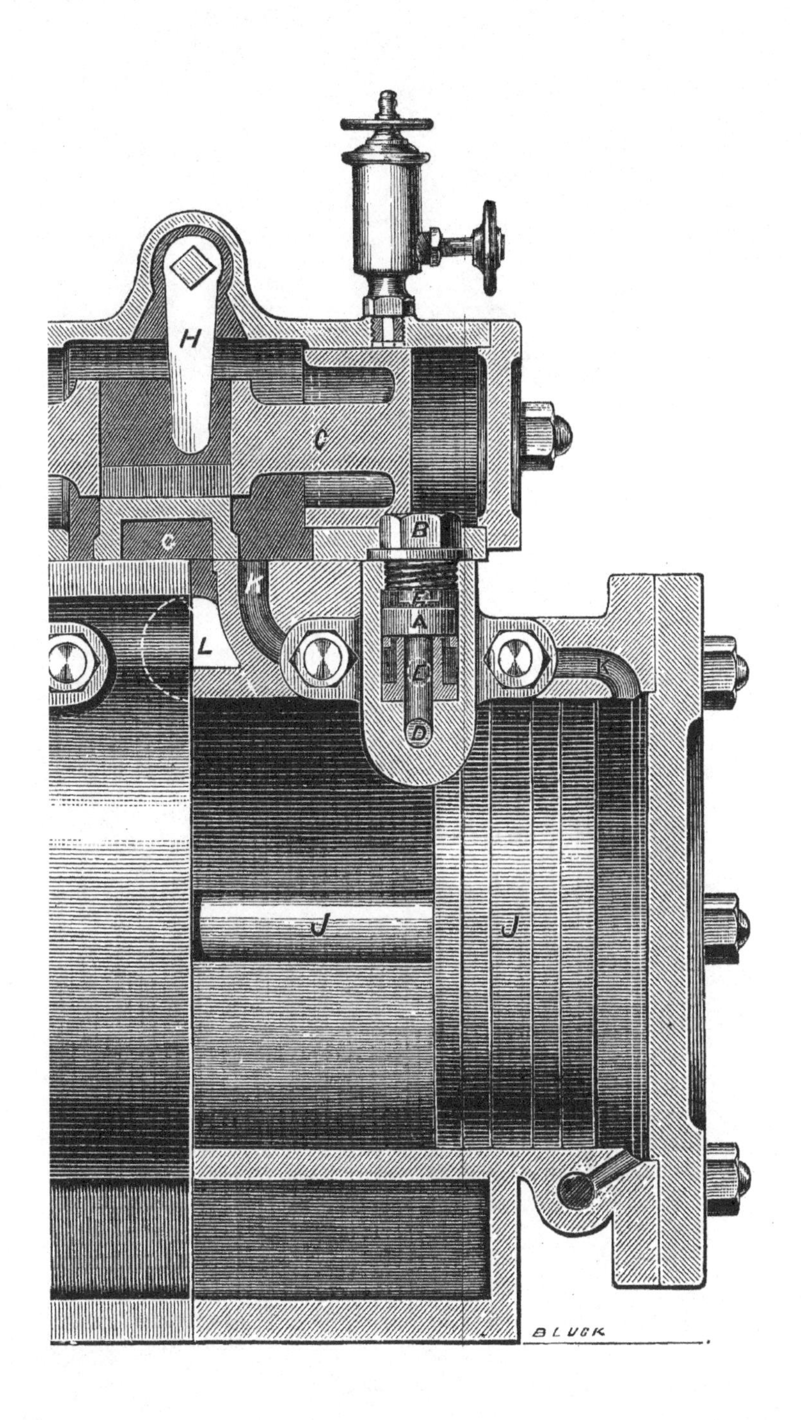
H
C
B
C
K
F
A
L
E
D
K
J
J
BLUCK

the auxiliary-valve, and regulates its action; C C are the plungers which actuate the main-valve, G; D is a port leading from the interior of the cylinder to the underside of the auxiliary-valve, A; E is a port leading from the auxiliary-valve to the end of the plunger, C, whereby its position is changed; F is a port through which the steam passes to the top of the auxiliary-valve, replacing it in its seat; G is the main-valve, an ordinary flat-valve moved mechanically by the plungers C C; H is a lever worked by the starting-handle; J J are the piston and piston-rod; K is a steam-port; L is the exhaust.

The makers of this pump, Messrs. The Imperial Steam-Pump and Engineering Company (Smethwick, near Birmingham), claim the following advantages for it:

'For simplicity of construction, economy of steam, certainty of action, and durability, the "Imperial" steam-pump cannot be excelled. All the working parts of the engine are under the control of the starting-handle. The action of the pump is obtained without the use of extraneous gear or tappets; their function is performed by small auxiliary-valves, external to the cylinder, and operated only by the steam. The whole of the working parts are easily accessible.' *

THE 'STANDARD' PATENT STEAM-PUMP.

This pump was introduced about eight years ago. The main steam-cylinder has mounted upon it a cylindrical valve-chamber containing an annular slide-valve with piston-ends. The slide-valve is actuated by steam from the main-inlet, through the medium of a cylindrical leading-valve traversing in a small chamber across the centre of the main-valve.

'Instead of having valves at each end of the steam-cylinder, as in other steam-pumps, which are actuated by

* Since these papers have gone to the press an improvement has been made in the valve-gear of the 'Imperial' pump. The steam is held on the top of the small side valves during the whole of the stroke, instead of only for a short time at the beginning, by which means the action of the gear is made more certain.

the bumping of the piston against them, rendering them very liable to derangement, there are in the "Standard" pump two additional steam port-ways, so that when the piston moves on its stroke beyond one of these ports, the steam passes up it and moves a small piece, which directs the steam to either the one or the other end of the slide-valve.

'There is absolutely nothing in the valve arrangement to get out of order, which is a most important point in these pumps. The "Standard" has a long stroke, and works with great effect' (Letter from the maker).

'It is claimed for the arrangement of the ports that it obviates entirely the dead-centre which occurs in other systems of actuating the slide-valve. The pump will start at any portion of the stroke. Every part can be easily got at, and the pump can be worked by a labourer of ordinary intelligence. Many of these pumps are working under water 50 feet, ... a feature which rendered it specially valuable, as the fact of a mine or pit being flooded would not interfere with the working of the pump.'*

Mr. Thomas A. Warrington, 30 King Street, Cheapside, London, E.C., is the maker of this pump.

TURNER'S 'TAPPET' STEAM-PUMP.

Mr. W. Turner has made engines of the type shown at fig. 70 for The Clifton Kersley Coal Company and Messrs. Andrew Knowles and Sons, of the following respective sizes: 34-inch steam-cylinder, 9-inch double-acting ram, and 6-feet stroke; 24-inch steam-cylinder, 10-inch single-acting ram and 6-feet stroke. These engines are fitted with the Cornish valve-gear.

The above engine is peculiarly applicable to deep mines, and for raising large quantities of water. Its action is steady and reliable, and may be described as follows:—

'On steam being admitted into the cylinder, the small roller, R, advances until it comes in contact with the tappet, T, which on being depressed relieves a small catch, C, and

* *Royal Cornwall Polytechnic Society's Report for* 1875, p. 67.

Fig. 70

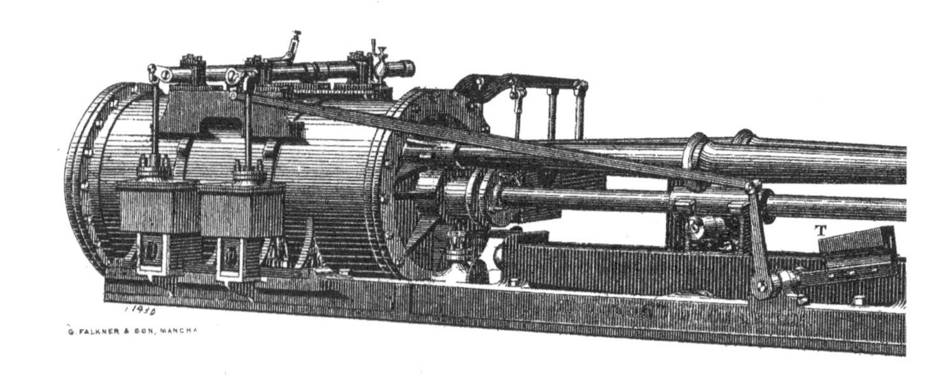

Fig. 76

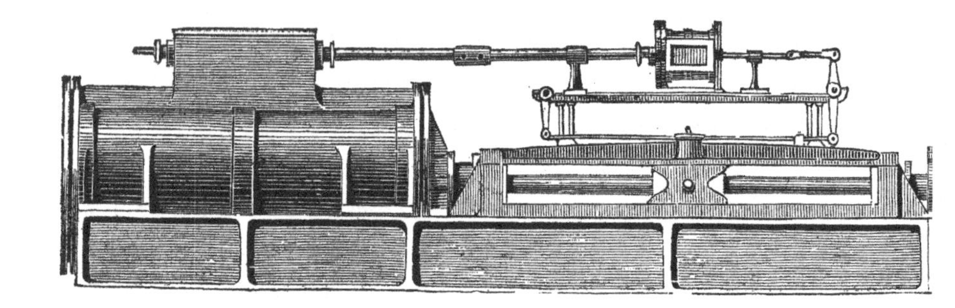

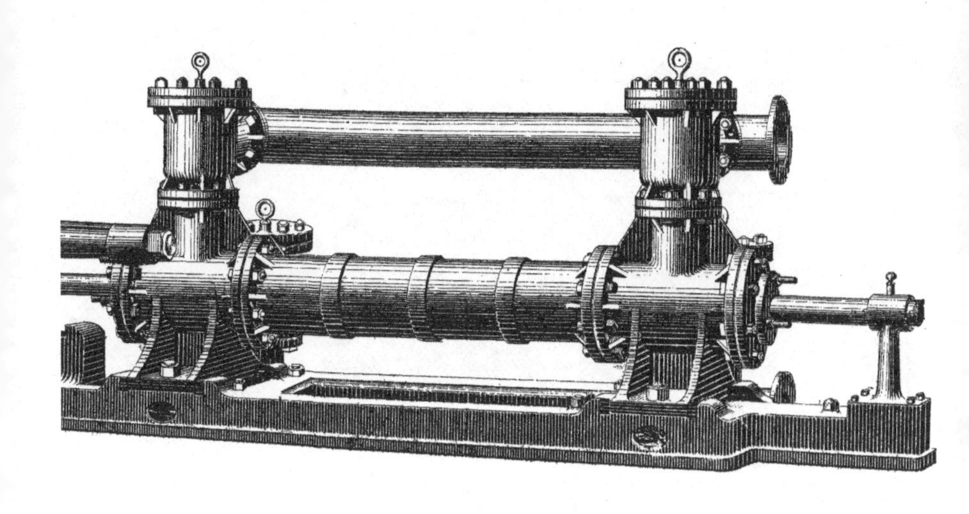

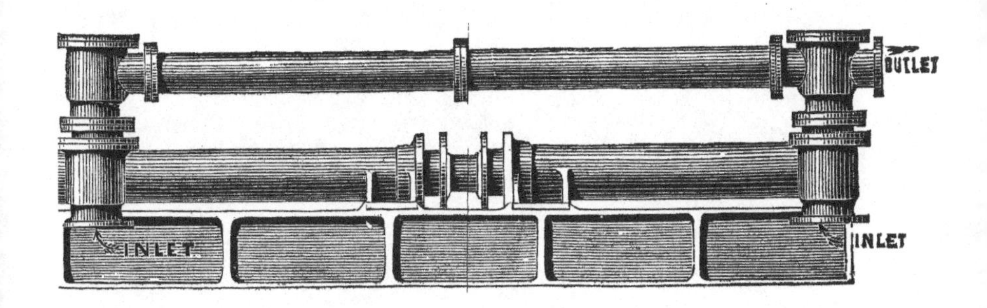
OUTLET
INLET
INLET

setting free a weight, reverses the action of the valves, and prepares the engine for the return stroke.

'William Turner has patterns for various sizes, and is prepared to adapt them to any condition.

'When applying for pumping engines for mines, water-works, &c., it is necessary to state the vertical height from the surface of the water from which the pump obtains its supply, and the point at which it is to be delivered, also the number of gallons or cubic feet required per minute. The pump should never be more than 25 feet vertically above the source of supply, in which case a foot valve should be applied.' (Maker's remarks.)

2. SINGLE STEAM-CYLINDER, EXPANSIVE, NON-CONDENSING ENGINES, WORKING DOUBLE-ACTING PUMP.

Although classed as expansive, these engines, when used underground for deep drainage, are only expansive to a small extent. Without a high initial pressure of steam—a moving mass much greater than the mean water-load—to absorb the excess of pressure at the beginning of the stroke and a great piston speed, or, in the absence of a great moving weight, a high initial pressure of steam and a great speed, it is impossible to get any appreciable expansion from engines of this class. The steam can only be expanded down to the water-load, but in no case below the pressure of the atmosphere. If, for instance, the steam be supplied at 60 lbs. per square inch and the water-load be 20 lbs., the steam may be expanded to about 22 or 25 lbs. per square inch, which are then lost. It will thus be seen that, although expansion is possible in these engines, it is always attended with a comparatively great loss of steam. If a condenser be applied, the steam could be expanded below the water-load; the extent to which this could be done depending upon the value of the vacuum and the friction of the machinery and water.

Davey's Patent 'Differential' Engine.

This engine works a double-acting ram-pump, as seen in fig. 71, and is specially adapted to forcing great heights and pumping water containing grit or mud. As has been already remarked, a pump for a permanent underground engine should have no internal packing or wearing surface which may be affected by grit or sand in the water. The pump should have a plunger and external stuffing-boxes such as are used in connection with this engine. Any leakage at once manifests itself, but when a piston is employed much water may be passing it at every stroke without detection. This pump consists of two working cylinders (or ram-pumps) fixed in centre line of the steam-cylinder, with a long ram attached to the piston-rod, and working through a stuffing-box at each of the inner ends of the pump-cylinders.

Fig. 71.

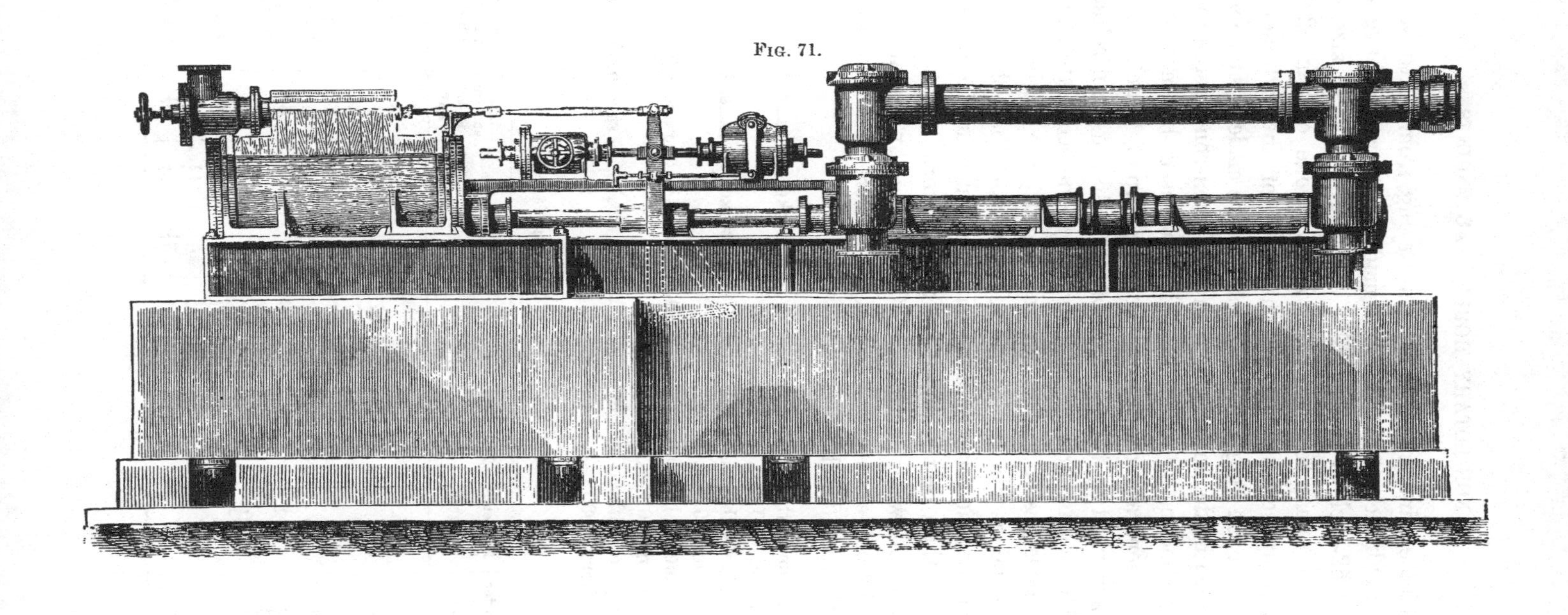

This engine is made in sizes varying from 10 to 40-inch steam-cylinders; the largest made appears to have been one of 52-inch cylinder. The pump-rams vary in size from 4 to 12 inches in diameter, with a stroke of from 2 to 5 feet. Any combination in the sizes of the cylinder and rams and the length of the stroke can be made. This engine has been made both condensing and non-condensing; when made condensing the steam is condensed by means of a patent separate condenser, the air-pump of which is sometimes worked by the pressure of water in the main column.

The engine and pump stand upon one bed-plate; in one of the designs of the condensing engine, the engine, pump, and condenser are placed upon the same bed-plate. The type of engine for pumping gritty water is much more expensive than that required for clean water.

The following particulars of work done by engines of this class in collieries are supplied by Messrs. Hathorn, Davey, and Co., Leeds, the makers:

One engine with 12-inch cylinder, 15-inch stroke, to raise 6,000 gallons per hour 300 feet.

One engine with 14-inch cylinder, 20-inch stroke, to raise 8,000 gallons per hour 190 feet.

One engine with 14-inch cylinder, 20-inch stroke, to raise 8,800 gallons per hour 320 feet.

Two engines with 20-inch cylinders, 36-inch stroke, to raise 6,000 gallons per hour 1,000 feet.

Table of Sizes, Capacities, &c.

Diameter of steam cylinder	Diameter of plunger	Length of stroke	Gallons per hour	Dia. of exhaust pipes	Dia. of steam pipe	Dia. of suction & delivery pipes	Height water can be raised with 40 lbs. boiler pressure
in.	in.	in.		in.	in.	in.	ft.
10	4	36	3,837	2½	1½	3	400
12	4	36	3,837	2¾	2	3	575
14	4	36	3,837	3	2½	3	778
16	4	36	3,837	3½	2¾	3	1,024
18	4½	36	4,860	4	3	3½	1,015
20	5	48	6,997	4½	3¼	4	1,019
22	5½	48	9,097	5	3½	4¾	1,020
24	6	48	10,080	5½	4	5	1,027
26	7	48	13,692	6	4¼	5¾	870
28	7	60	15,648	6¼	4½	6¼	1,010

THE 'SELF-GOVERNING' ENGINE.

The self-governing motion of the valve-gear, which is the peculiar feature of this engine, is the invention of Messrs. Cope and Maxwell. The engine is constructed in two forms: one in which both the cataract and the valve-moving cylinder travel, and in the other the valve-moving cylinder is stationary and only the cataract-cylinder travels. 'The self-governing motion was introduced to ensure economy of fuel, to procure a positive motion of the valves at all times, and for regulating the motion of the main-piston under all circumstances to which pumping engines are subject by suddenly increased or diminished load on the pump or pumps, and thereby render the engine safe under variable loads, by varying the pressure of steam in the main steam-cylinder or cylinders, according to the load or resistance against which the piston works. This is accomplished by actuating the main slide-valve of the engine by the combined motion of a valve-moving cylinder and a valve-moving piston. The valve-moving cylinder receives its motion from the main-piston of the engine by means of levers, and consequently has a motion which varies with the velocity of the main-piston. The valve-moving piston is actuated by the direct pressure of steam, and has its motion controlled and rendered constant by a cataract-governor. The valve-moving cylinder and valve-moving piston travel in opposite directions, the one acting upon the main-valve to cause it to open the steam-ports, and the other to shut off the admission of steam. The resultant of these two motions is to cause a motion of the main slide-valve, which varies with every minute increase or decrease of the resistance of load upon the main-piston.'*

In the arrangement of the valve-gear to which the accompanying illustrations and the following description refer, both the valve-moving cylinder and the cataract-cylinder (or go-

* *Direct-acting Pumping Engines*, by Mr. P. R. Björling, pp. 84 and 85, a paper read before the Society of Engineers, May 7. 1877.

I 2

vernor), travel. Fig. 72 is a perspective elevation, fig. 73 a longitudinal section, fig. 74 a plan, and fig. 75 a transverse elevation of an engine with a double-acting ram-pump. It consists of one main horizontal cylinder, mounted upon a cast-iron bed-plate, with steam and exhaust-ports, and D slide-valve of the ordinary type; one steam starting-cylinder, and one cataract-cylinder, mounted upon an entablature in the centre line of the main slide-valve spindle, the piston-rods of which are a continuation of the same. The special feature of this machine is the self-governing valve arrangement. A is the main-cylinder, provided with the usual steam and exhaust-ports and passages; B, the main-piston; C, the valve-chest; D is the main slide-valve of the ordinary D type. E is the valve-spindle, which is prolonged, and also acts as the rod for the valve-moving piston, F, and the cataract-piston, G; Q is the valve-moving cylinder, on the top of which is a small valve-chest, H, in which works a supplementary valve, I; this valve admits steam alternately on either side of the valve-moving piston, F. N is the supplementary valve-spindle, which is provided with two tappets, O; and on the end of this spindle provision is made for the starting-handle, shown in plan, fig. 74; J is a pipe admitting steam from the main valve-chest to the valve-moving cylinder-chest, and K is a pipe leading from the valve-moving cylinder to the main-exhaust, L; these pipes work in glands, shown in the plan. The cataract-governor, P, consists of a plain cylinder, with a port or passage leading from one end of the cylinder to the other; and an adjusting plug is fitted in the middle of this passage, for regulating the speed of the valve-moving piston, and, consequently, the main slide-valve, which is attached to the same rod; R is the main-levers, working on a fulcrum, S. The bottom ends of these levers are. by the two links, T, connected to a horn or arm, U, which is fastened to the main piston-rod, V, with two set-screws or a cotter. At the top part of the main-levers, a little distance above the fulcrum, are two slide-links, W, which connect the levers to the cataract-cylinder. The top ends of these levers are prolonged, and fitted with small rollers for working the

Fig. 72.

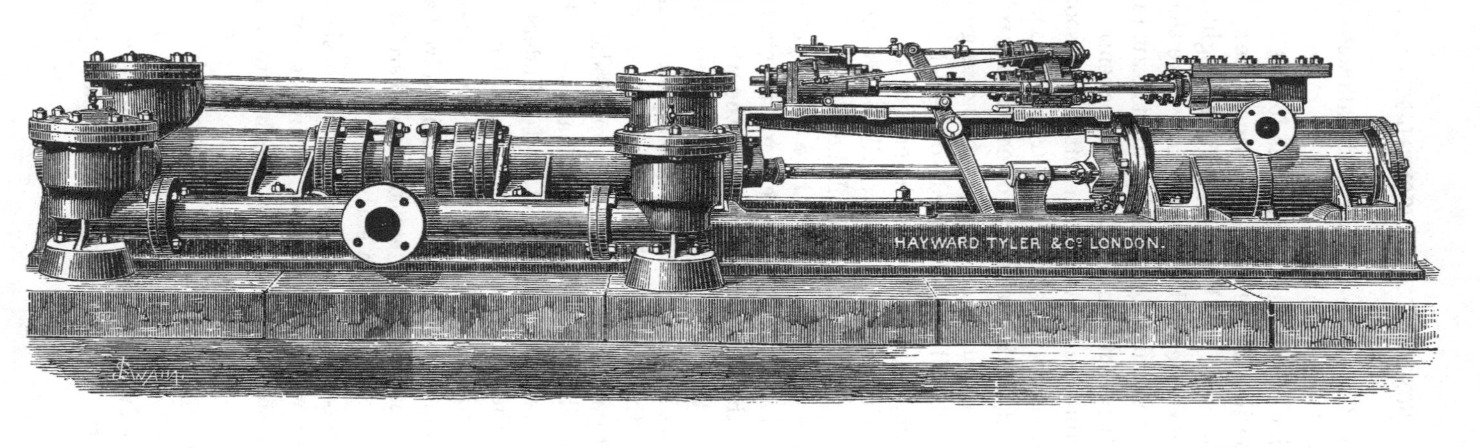

tappets, O. The valve-moving and cataract-cylinders are coupled together by two stays, X X.

The action of this valve-gear is as follows:

By the position of the valve, I, shown on the sectional elevation, fig. 73, steam will be admitted to the right-hand side of the valve-moving cylinder Q, and thus move the main-valve towards the left, and by that means admit steam into the main-cylinder at the right-hand end, and consequently commence to move the main-piston towards the left; but as soon as the piston commences to travel the top part of the main-levers, R, will also travel, but in the reverse direction (towards the right), and the valve-moving and cataract-cylinders go with it; the slide-valve will still move onward to the left, till the valve-moving piston, F, has come to the end of the valve-moving cylinder, Q, when the main-valve will be taken back by the valve-moving cylinder, and cut off the steam. The rollers on the top of the main-levers will strike the tappet, O, before the main-piston has arrived at the end of its stroke, and thus by reversing the supplementary valve the whole is reversed. Now, suppose that the engine has commenced its stroke, and the resistance is very great, the main-piston will travel on slowly, but the valve-moving piston will travel as fast as the cataract-governor will allow; the main slide-valve will go to the end of its stroke, before the main-levers have time to pull the valve-moving and cataract-cylinders back, thus cutting off the steam late, but if the load be light, the main-piston will travel fast, and pull the valve-moving and cataract-cylinders back, before the main-valve has had time to travel to the end of its stroke, and therefore cut off the steam from the main-piston early.

The water-pump is double-acting, consisting of two single-acting working-pumps or cylinders in a line with the main piston-rod, which is attached to a long hollow ram working through a stuffing-box at one end of each working-pump. There is an inlet-branch in the side of each cylinder near the outer end, to which is connected the suction valve-chest (or nozzle). The delivery valve-chest stands directly over the same end of each cylinder, and at

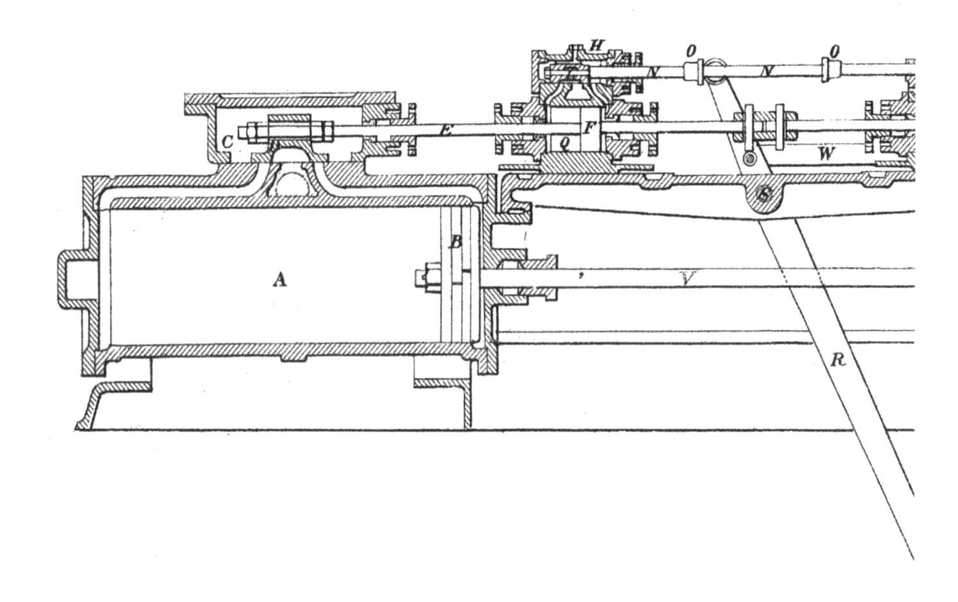

FIG. 74.

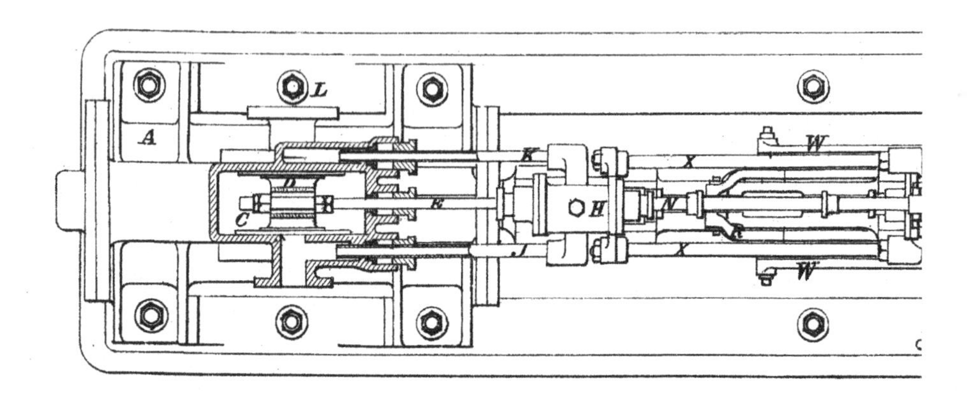

Fig. 73.

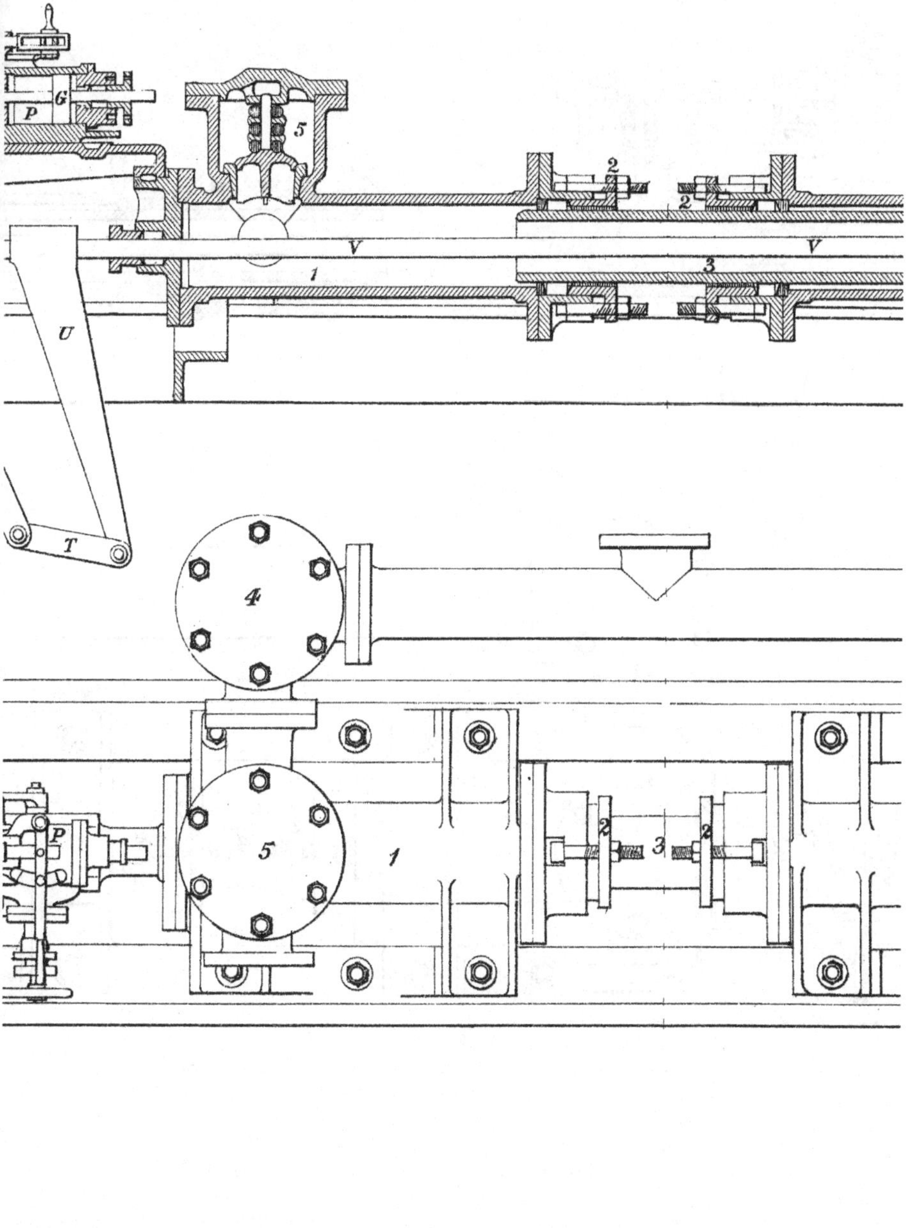

Fig. 73.

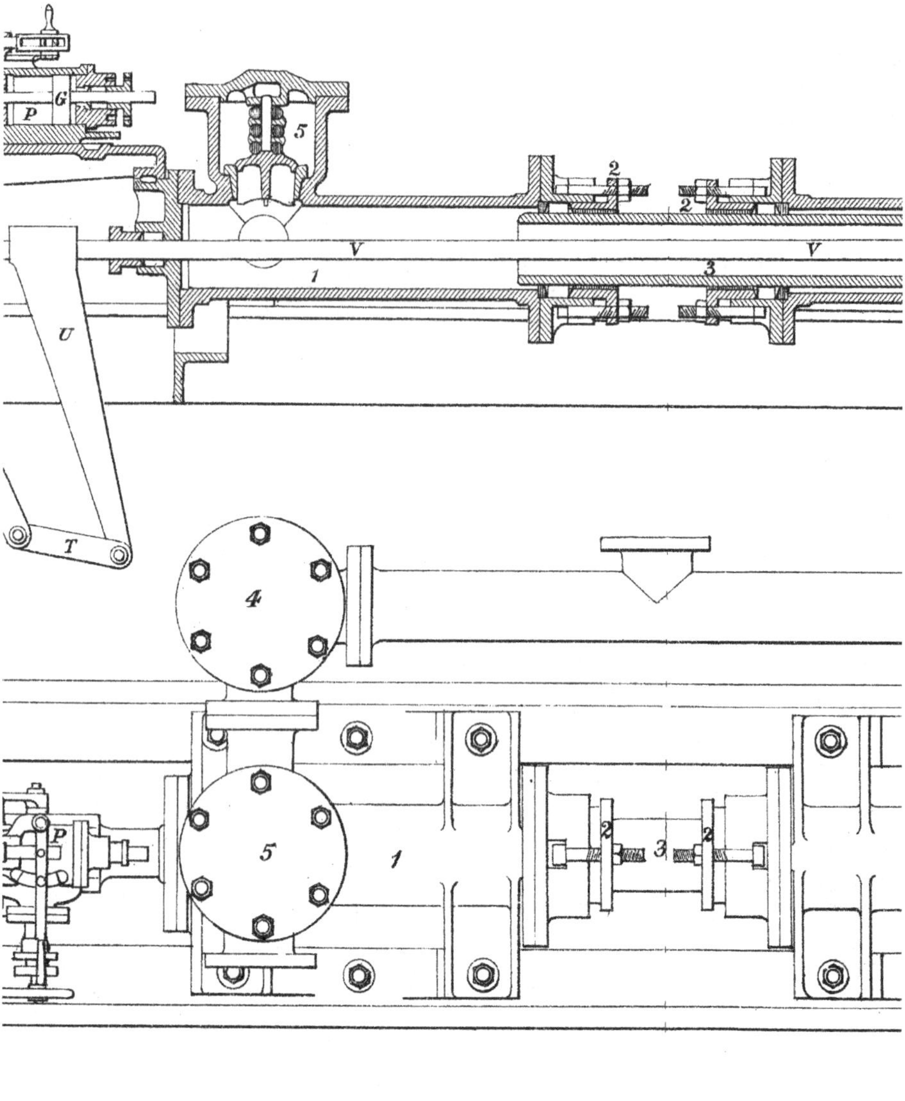

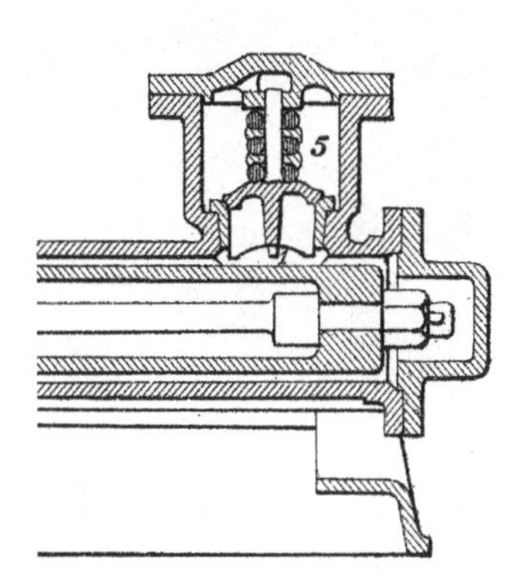

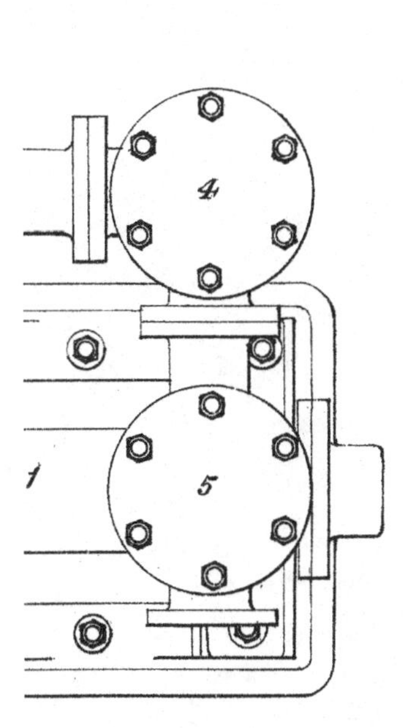

Fig. 75.

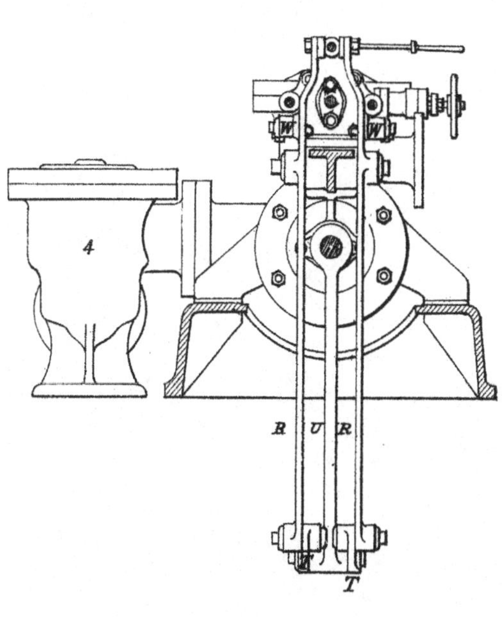

right angles to the inlet-branch. The drawing represents metallic wing-valves in the delivery-chest, with india-rubber springs for controlling the motion of the valve and giving it a quick return.

'When this engine was exhibited at the Royal Agricultural Society's Show at Birmingham last year (1876), the delivery-pipe was throttled by means of a screw-down valve, until the resistance on the pump-ram was so great as to nearly balance the pressure of steam in the cylinder, of course materially reducing the speed of the piston. The delivery-pipe was suddenly opened by two sharp turns of the hand-wheel when the main-piston had completed three-quarters of its stroke, without any appreciable augmentation in the speed of piston to which the cataract-governor was adjusted. The author thinks this was a good proof of the sensitiveness of the self-governing gear. The slide-valve of this engine works in a similar manner to an eccentric-moved valve, opening and cutting off steam quickly, not, as is the case in most direct-acting automatic expansive engines, with a slow motion and "wire-drawing" the steam.

'Another great advantage in this gear is the ease with which the valves are set; almost any ordinary engine-fitter can do it. . . . These engines can, of course, be made either high or low-pressure, compound or non-compound, according to the degree of economy of fuel required, the space which can be allotted to it, and other circumstances.' *

Messrs. Hayward Tyler and Co., of London and Luton, are the sole makers of this engine in this country.

Turner's 'Tappet Expansion' Engine.

'Fig. (76) illustrates a very simple tappet-engine. The whole of the gear being external it only requires an ordinary mechanic to keep the machine in working order. The tappet-gear reverses a small balance-valve of the subsidiary-cylinder, and this in turn reverses the main-valve. Its action may be

* *Direct-acting Pumping Engines*, by Mr. P. R. Björling, pp. 86 and 87.

regulated by a cataract, or the exhaust may be throttled. The pump is of the double-ram type, with external packing-glands. A small amount of expansion may be obtained by having a large lap on the valve.

'The desire for cheap pumping engines has led to the introduction of many systems, the most simple being the tappet-engine, which works without expansion (as illustrated at fig. 70). I have made many engines of this class, which have now been working for some years with excellent results; they are very simple and very reliable, can be regulated to any required speed, and having few parts, the expense for maintenance is very small. The design above illustrates a tappet-engine, with apparatus for expansion, which may be varied and adjusted to suit the speed and stroke of the engine. The engines may be fitted with piston or double-ram pumps, and with suitable single or double-beat Cornish valves of cast-iron or gun-metal, as may be thought most desirable' (Maker's remarks).

The accompanying table contains particulars of sizes, capacities, and dimensions of this engine.

Diameter of cylinder	Diameter of ram	Stroke	Height	Delivery per hour	Space required
in.	in.	in.	ft.	gall.	ft.
18	6	36	500	7,920	18 × 3
24	7	36	700	10,800	19 × 3½
24	8	48	500	15,000	24 × 4
27	9	48	500	18,000	24 × 5
34	10	72	650	27,000	36 × 6
36	12	72	500	36,000	36 × 6
45	14	72	583	51,000	36 × 6

3. SINGLE STEAM-CYLINDER, EXPANSIVE, CONDENSING ENGINES, WORKING DOUBLE-ACTING PUMP.

Any of the engines noticed may be made condensing. All engines above a small size ought to be supplied with a condenser, as a great saving of steam is thereby effected. The vacuum which it produces would remove a back-pressure generally exceeding 7 or 8 lbs. per square inch, and sometimes as much as 12 or 13 lbs., depending upon the quality of the vacuum. The steam could be expanded to nearly that extent below the mean water-load, a small margin being necessary for the friction of the machinery and the water through the pipes. The classes of engines previously described are made without condensers. If the reader will refer to the tables of prices at page 46 he will observe that the condensers are extras. It is therefore appropriate to class them as non-condensing engines.

The only examples we shall give of engines of this class are those of Messrs. Davey and Parker and Weston.

DAVEY'S PATENT 'DIFFERENTIAL' ENGINE.

This is a condensing type of the engine at fig. 71.

The following are examples in use in collieries:

One engine, with 12-inch cylinder, 20-inch stroke, to raise 6,000 gallons per hour 250 feet.

One engine, with 19-inch cylinder, 20-inch stroke, to raise 15,000 gallons per hour 300 feet.

One engine, with 18-inch cylinder, 24-inch stroke, to raise 18,000 gallons per hour 135 feet.

One engine, with 22-inch cylinder, 24-inch stroke, to raise 12,000 gallons per hour 480 feet.

Two engines, with 26-inch cylinder, 36-inch stroke, to raise 78,000 gallons per hour 240 feet.

One engine, with 30-inch cylinder, 36-inch stroke, to raise 18,000 gallons per hour 480 feet.

Table of Sizes, Capacities, &c.

Dia. of steam-cyl.	Dia. of plunger	Length of stroke	Gallons per hour	Dia. of steam pipe	Dia. of suction and delivery pipes	Height water can be raised with 40 lbs. boiler pressure
in.	in.	in.		in.	in.	ft.
20	6	36	8,640	3½	4½	713
22	7	48	13,692	3½	6	627
24	6	48	10,089	4	5	1,027
26	7	48	13,692	4	6	876
28	9	60	25,920	4½	8	615
30	7	60	15,650	5	6½	1,150
34	10	60	32,000	5½	9	737
36	9	72	29,200	6	8½	1,017
40	8	72	23,000	6½	7½	1,600
50	12	72	48,210	8	11	1,105

Parker and Weston's Patent Engine.

Figs. 77 to 88 are illustrations of the details of Messrs. Parker and Weston's double-acting ram-pump, with condenser, manufactured by Messrs. The Coalbrookdale Co. Figs. 77, 78, and 79 are a side elevation, plan, and an end elevation

Fig. 85.

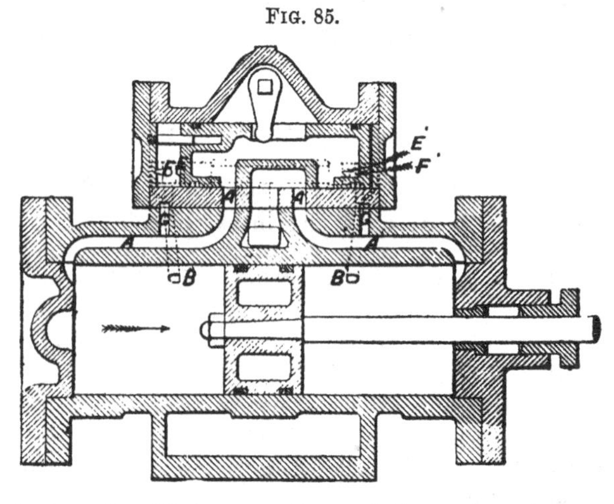

respectively; fig. 80 is a section of the cataract-governor; fig. 81 is a cross-section of the pump; fig. 82 is a longitudinal section of the engine and pump; fig. 83 is a cross-section of the engine; fig. 84 is an enlarged half plan and section of

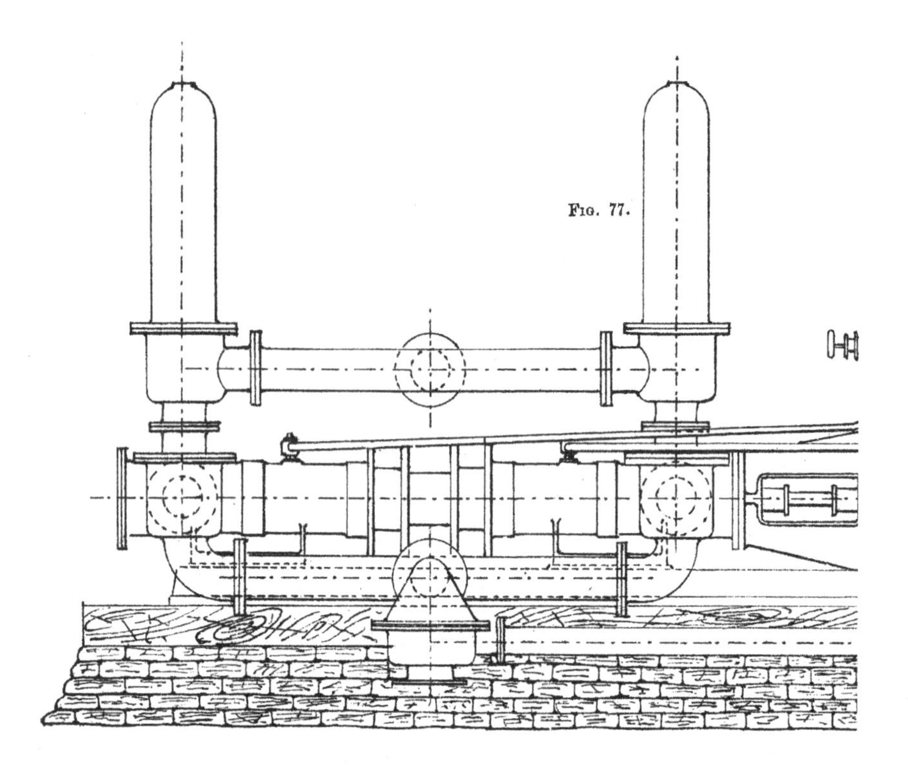

Fig. 77.

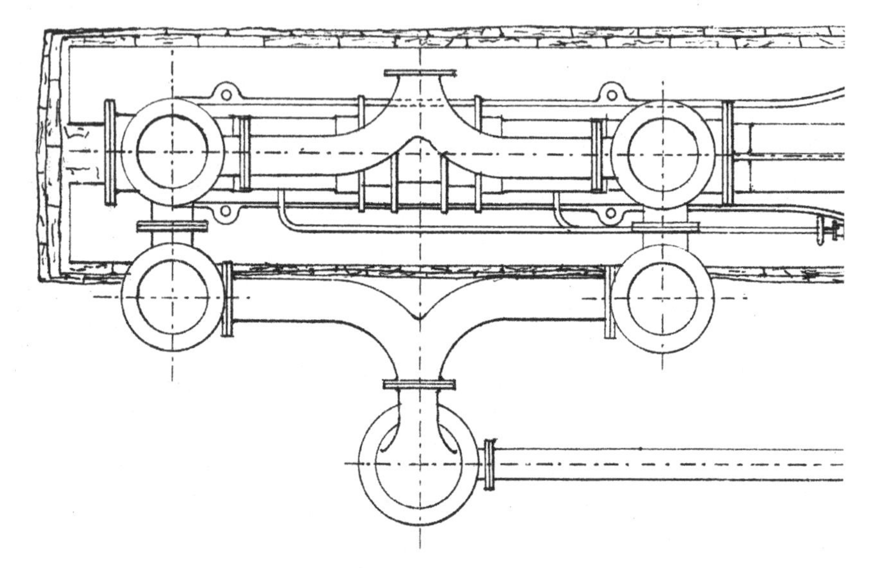

Fig. 78.

Fig. 79.

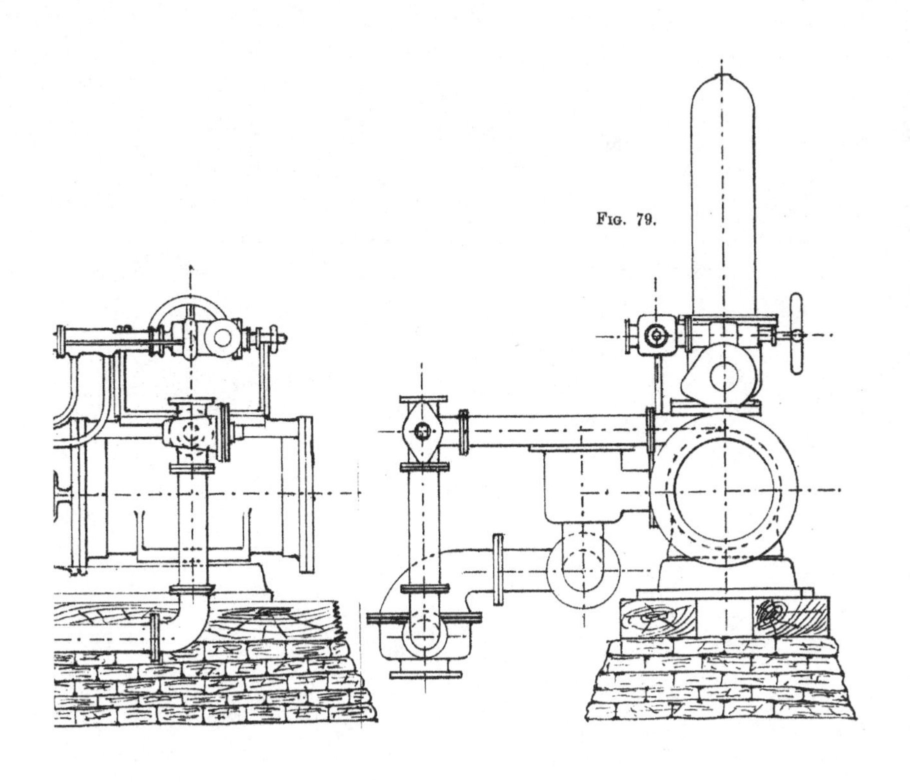

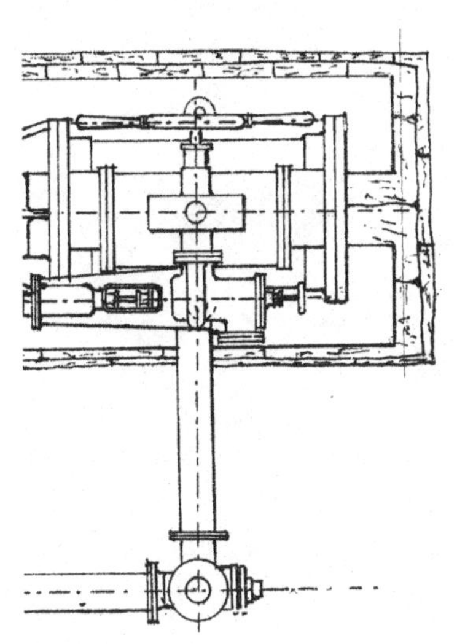

RAM PUMP with Cataract and Condenser

Scale ½" = 1 foot

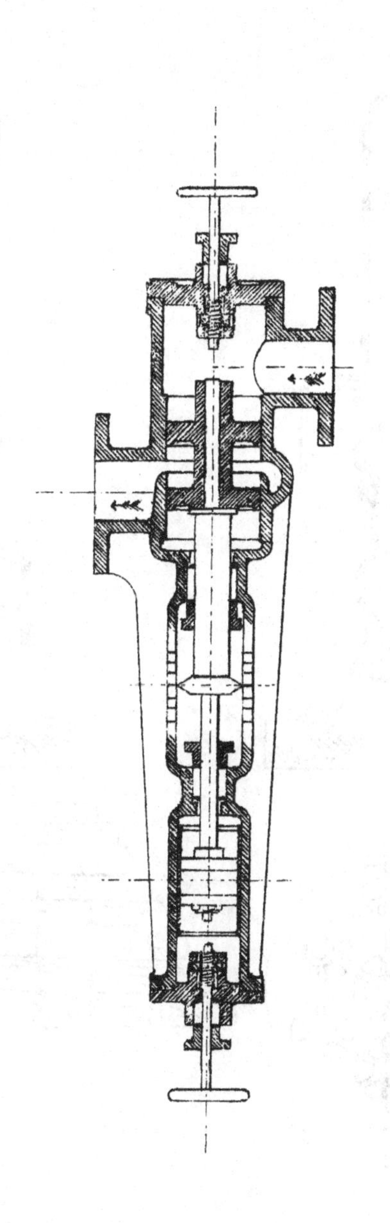

Fig. 81.

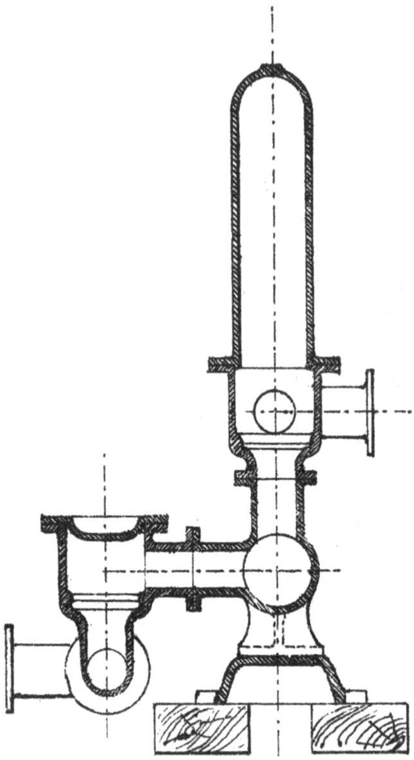

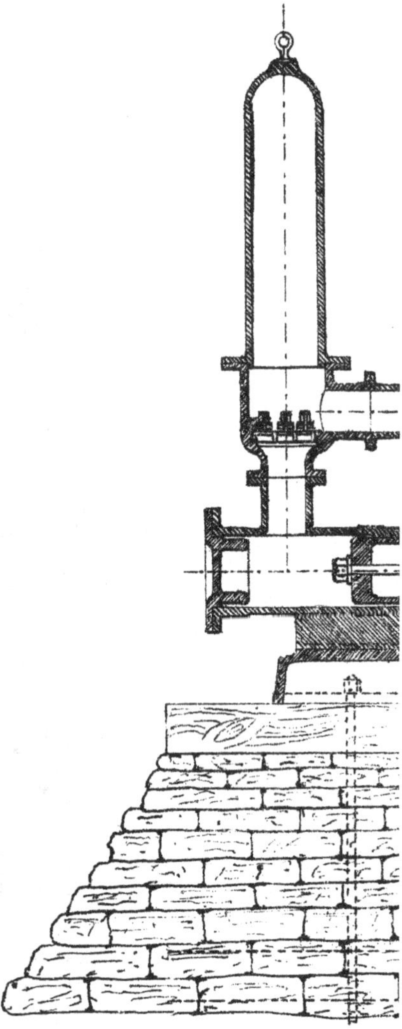

Fig. 84.

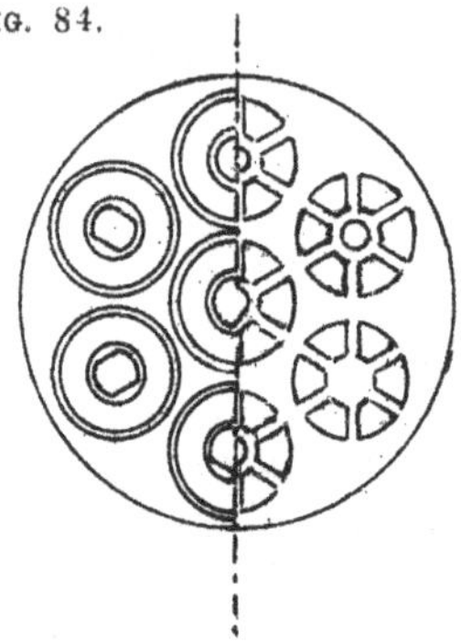

Fig. 82.

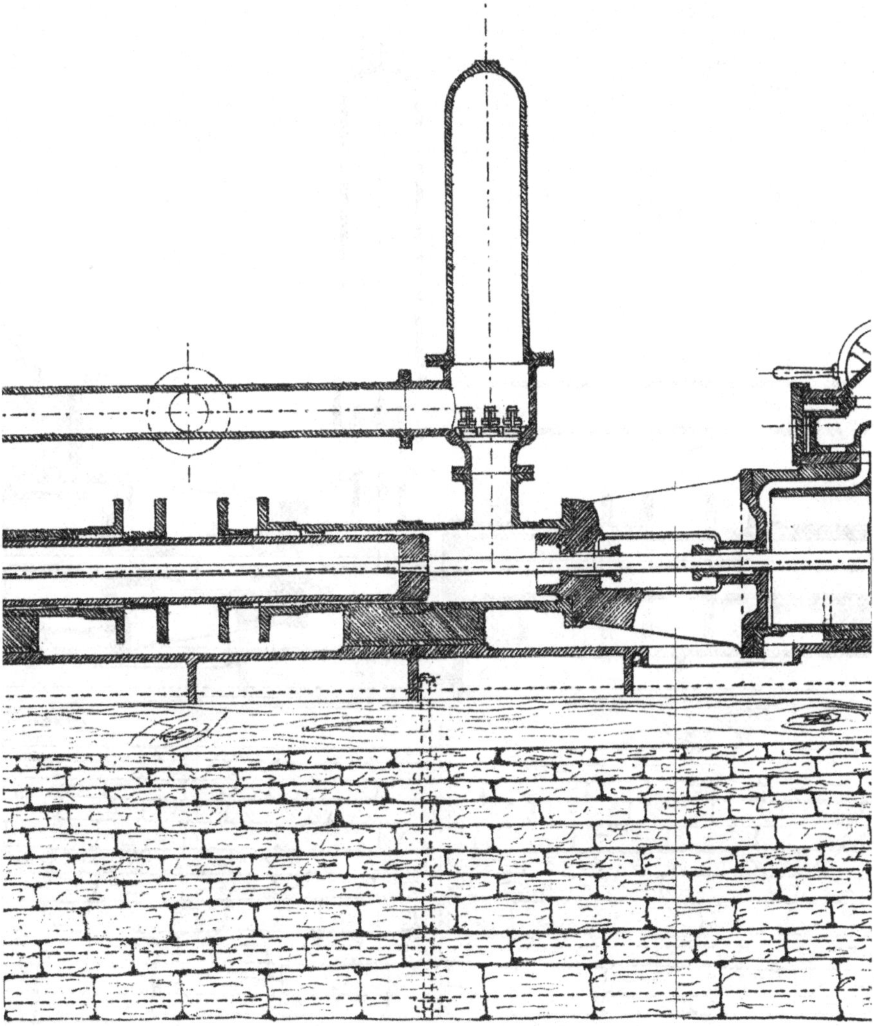

Fig. 82.

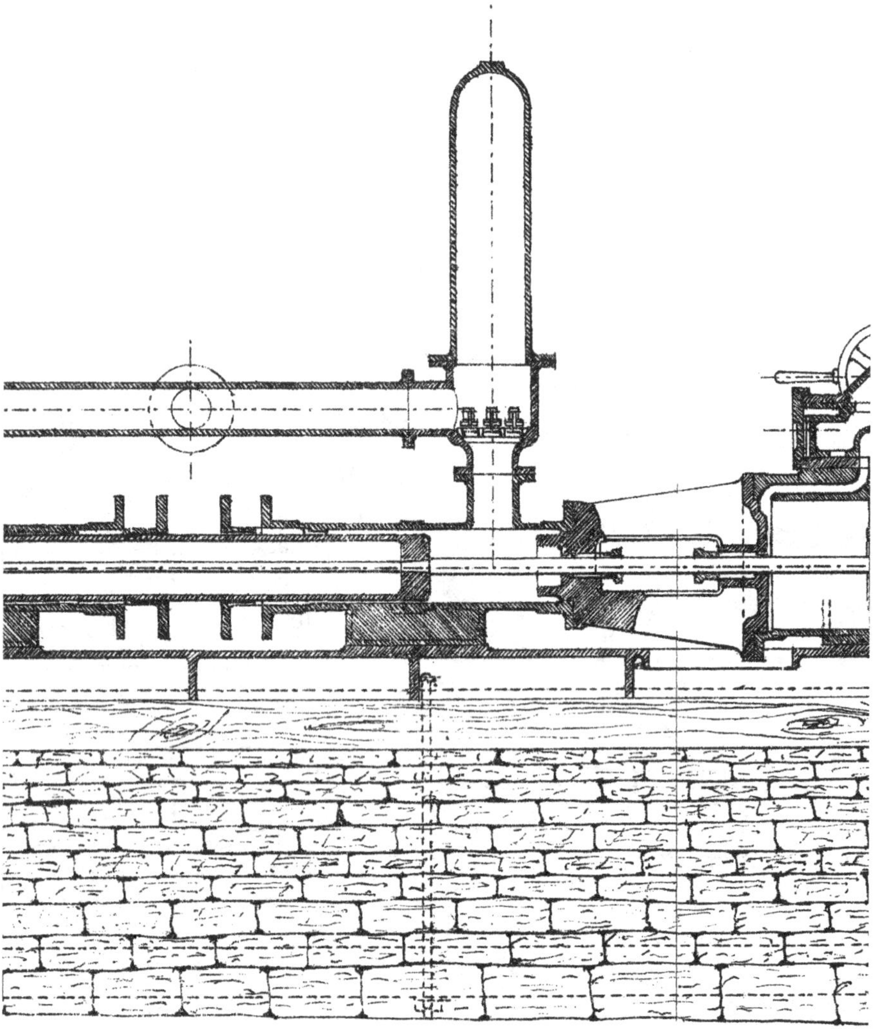

Fig. 83.

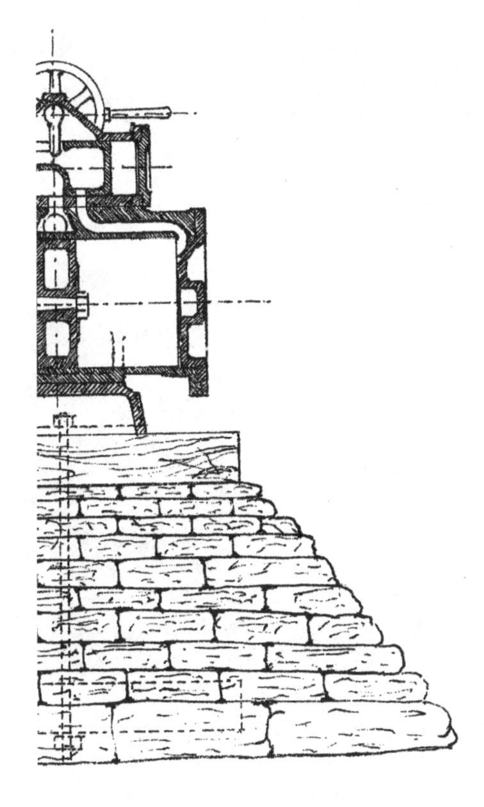

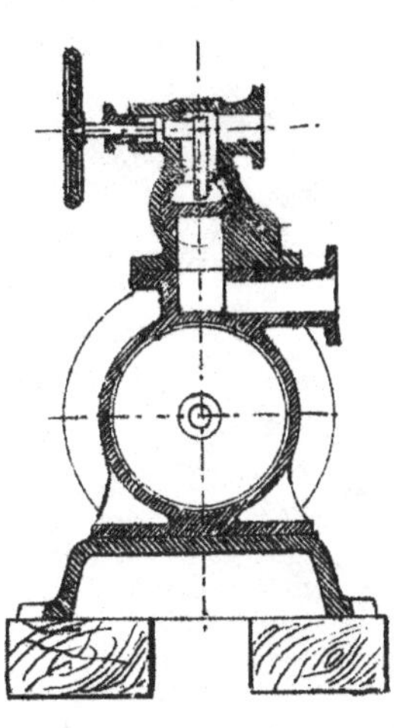

RAM PUMP

Scale ½" = 1 foot

the pump-valves; fig. 85 is a sectional elevation of the steam-cylinder and valve-chest; fig. 86 is a plan showing the steam-chest in section; fig. 87 is a cross-section of the steam-cylinder and valve-chest; and fig. 88 a plan

FIG. 86.

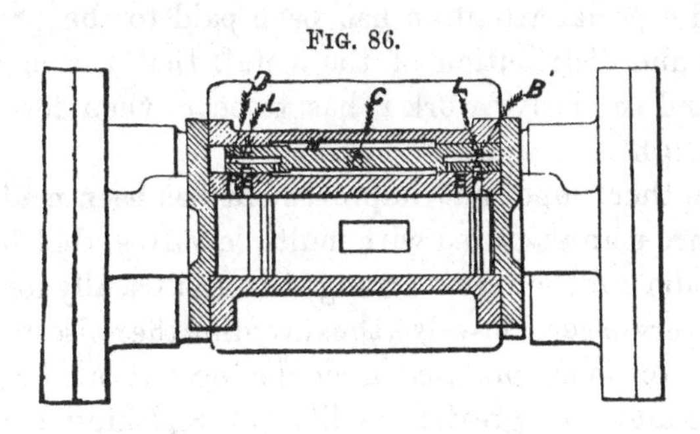

of the valve-face. The author is indebted to the makers for the following description: 'It will be seen on reference that the steam-cylinder and pumps are mounted on a cast-iron bed-plate, so that the engine may be

FIG. 87. FIG. 88.

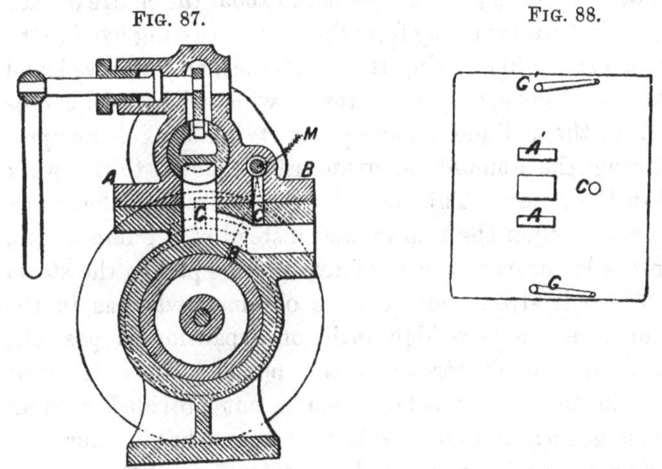

safely and easily placed on the roughest ground, a convenience that will be fully appreciated by mining engineers, who, perhaps in a case of emergency, have had to prepare costly foundations for an ordinary pump, and have lost most valuable time. Attention has been paid to the design that it may look neat, and also that it may possess portability; the valve-boxes, pipes, and pump-barrels are made separate, so that they may be detached and moved easily. All parts subjected to great strain are made cylindrical, and especial attention has been paid to the proper proportion and distribution of the metal, that the engine may withstand the heavy work it has to bear when forcing to a great height.

'Another important improvement has been made: these pumps are manufactured with multiple-valves, that is, several small india-rubber valves arranged concentrically for forming one delivery or suction-valve, thus avoiding the noise inseparable from the ordinary practice of employing but one large valve, and also allowing greater facility for replacing a damaged one. These multiple-valves work with comparatively small lift, which enables them to close with great precision, a virtue not possessed by the ordinary sort when used for large pumps. There will also be seen in the drawings (figs. 77 and 78) two small wrought-iron pipes connected to about the centre of each pump-barrel, and running from thence to a casting fixed to the steam-chest. This casting is the cataract, an improved and most efficient governor, by means of which the steam can be cut off in the cylinder at any point of the stroke, thus proportioning the amount of steam used exactly to the work done in the pump. This desirable end is accomplished more successfully when the ram is used instead of a pump-piston, for in the latter case a ratio of expansion equal to the steam cut off at half stroke only can be obtained, whereas in the case of a ram a very high ratio of expansion is possible, equal to the whole stroke of the pump. There is great danger in the use of a large pump not provided with an efficient governor, and engineers know what a source of apprehension such an incomplete engine is, and the amount of attention necessary, for should such a pump miss its water by the failure of a suction-valve, the ram is driven back at the next stroke with tremendous force, in fact, with all the force of the steam in the cylinder, and the consequence is a smash of the pump or cylinder-cover, and a general disaster.

Now the cataract effectually prevents this, for as soon as the water pressure fails in the pipes leading to it, the steam is cut off, and no harm ensues. Thus it is seen that whatever the pressure in the pump the amount of steam is exactly proportioned thereto, and any sudden loads or failures of the valves are allowed for by the governor. One of the main points in connection with these direct-acting pumping-engines is the arrangement of valve or valves controlling the admission to, and the release of the steam from, the cylinder. In the present example the valves are entirely steam-moved, dispensing with tappets or other mechanical appliances likely to become deranged, and by a reference to figs. (85) to (88) the working of the valves will be understood. The main-valve is a hollow cylindrical casting, and the steam is admitted to the interior by an oblong hole in the top of the valve. The interior of the steam-chest is bored truly cylindrical, and the valve turned to fit, rings being provided near the end of the valve to prevent the passage of the steam. The auxiliary-valve which regulates the admission and release of the steam to move the main-valve is usually made of steel, and is of small diameter compared with the other. It works in a circular chamber horizontal and parallel to the main-valve. Both valves may be easily taken out and examined by removing one of the steam-chest lids.

'We will now endeavour to explain the working of the engine. The action of the steam-valves will be better understood by reference to figs. (85) to (88), which represent a steam-cylinder as applied to the smaller sizes of pumps. We will suppose the engine to have been started, and the piston to be travelling in the direction of the arrow. The main-valve is at the right-hand end of the steam-chest, and steam is entering the left-hand end of the cylinder by the port A. The auxiliary-valve is also at the right-hand end of the steam-chest. This state of things continues until the main steam-piston has passed a hole, B′, in the steam-cylinder, when the steam that has been used in propelling the piston rushes up the hole, B′, which communicates with the circular groove, L′, turned in the auxiliary-valve, passes through the cross passage, F′, and getting between the main-valve and the steam-chest lid, drives the valve over to the left, and allows the steam in the chest to find its way on to the right-hand end of the main-piston, and to exhaust from the opposite end. The piston is thus instantly stopped in its motion, and proceeds to make a stroke in the opposite direction. The instant the port A′ is charged with steam from the chest by the movement of the main-valve, a small hole, G′, conducts steam from the port A′ to the end of the auxiliary-valve, which is thus shot over to the opposite end of the valve-chest, and placed in a position for admitting and exhausting steam to and from the main-valve. It will thus be seen that the main-valve makes its stroke first, and on steam being admitted to the cylinder the auxiliary-valve then moves over. The steam used in moving the main-valve is exhausted through the passage, E′, thence into the groove, L′, in the auxiliary-valve, then by a number of small holes into a passage formed in the auxiliary-valve, and finally escapes into the chamber M, and thence by the hole C into the main exhaust-pipe. When the main-valve has made about half its stroke, it closes the outlet for the exhaust C, and the remaining steam is confined between the end of the valve and the steam-chest lid, and thus effectually prevents the valve from striking the ends.

'We will now explain the working of the cataract-governor, and a reference to the separate sectional drawing (fig. 80) will make the matter clear. The steam-valve consists of two equal pistons working in a cylinder, with a port between and steam-passages through them, although it will be noticed that they are not in equilibrium. To the left-hand is the water-cylinder, with a long ram working in it, and to which the water from the pump has free access on both sides of the ram, that is, the water is conducted from one end of the pump to the corresponding side of the piston in the cataract, and from the other end of the pump to the other side of the piston in the cataract. We have now the steam-valve cylinder and the water-cylinder, which are in one casting of the form seen in the drawings, and the steam-piston and pump-ram are connected by a piston-rod, which it will be seen is larger at one end than at the other. By making the piston-rod at the steam-valve end disproportionately large, a greater effective area is obtained on one side of the steam-piston than on the other end; by this means, as soon as the pressure fails on the right-hand of the pump-ram, the valves are driven to the extreme end of their travel, at the same time covering the port that gives admission of the steam to the valves of the engine. Again, should the pressure fail on the left-hand side of the pump-ram, the pressure on the other side immediately drives the valves to the other end, and so covers the port and cuts off the steam. So we see that immediately the water-pressure fails, no matter from what cause, the steam is cut off and the engine practically stopped. In starting the engine, when there is no pressure in the pump, it is obvious that the valves have cut off the steam; but the steam is not shut off entirely, as there is just a slight opening, which allows the steam to accumulate until there is enough to start the engine, when the pressure thus obtained will open the valves full. By means of a simple ball-valve on each pipe from the pump to the cataract, the time of cutting off may be varied, and any ratio of expansion is possible.' *

* The measure of expansion attainable in a single-cylinder engine without a heavy free weight in motion, or a great initial pressure of steam and an air vessel to absorb the excess of pressure at the beginning of the stroke, is very small. For a consideration of this subject see pages 47 to 52, 112, 121, and 128.

B. COMPOUND STEAM-PUMPS.

IN the non-rotary class of pumps which formed the subject of the remarks concluded at page 111, we observed that the piston was moved by steam at full pressure the whole length of the stroke, the steam being then wasted. It was evident that such a practice was incompatible with economy, and that the extension of this principle of pumping to engines of a large size, would incur such an extravagant waste of fuel as would be tolerated by no intelligent person, and which would not for a moment permit such engines to stand in competition with surface engines. The necessity of providing some better direct-acting appliances for pumping on a large scale than the simple steam-pumps did not want recognition from engineers, hence the extension of the principle to a combination of steam-cylinders by which the steam could be used expansively.

Without a considerable piston-speed and a heavy inert mass (which by, first, its inertia, and afterwards, its momentum, compensates for the diminution in the pressure of the steam) it is impossible to work underground single-cylinder engines expansively to an appreciable degree. The ratio of expansion that can practically be employed being dependent upon the amount of momentum that can be stored up in the moving mass during the first part of the stroke, and given out during the latter part, or, in the absence of a heavy moving mass, the excess of the initial pressure of the steam over the water-load, it is evident that engines of this class, unless rotary, can only be worked expansively to a very limited extent, and in any case there will be a great loss of steam, as it must be exhausted at a pressure exceeding that of the atmosphere. It will thus be apparent that, under the circumstances in which these engines are used, no device for making them expansive can prevent a comparatively great loss of steam, but the use of a condenser would permit the steam to be condensed at a considerably lower pressure.

A second cylinder to receive the high-pressure steam after it leaves the first cylinder becomes, therefore, a necessity for an economical ratio of expansion in underground engines.

In a prospectus of their compound pump the makers of one of the most extensively used simple pumps state thus: 'For confined situations, or where engines of a comparatively small size only are necessary, they (the simple pumps) will still meet all requirements, but their application will be very largely increased, since it has been found practicable to embrace the important features of expanding and condensing the steam, so that increased power may be obtained, and the consumption of fuel greatly economised.'

So far removed are these engines from the single-cylinder engines, in respect of economy, that their adoption in collieries may be suitable even when economy of fuel is a matter of some moment.

Among the best-known engines of this class are those of Messrs. Henry Davey (the 'Differential'), Parker and Weston, Cherry, Cope and Maxwell (the 'Self-Governing'), and Tangye Brothers and Holman. These machines are of a more imposing character than the simple pumps, being applied to more extensive drainage operations, and in their construction they are perhaps the most compact and the simplest forms of machinery for raising water which make provision for the expansion and condensation of the steam.

The low-pressure cylinder is usually about one and three-quarter times the diameter of the high-pressure cylinder; sometimes it is twice, and in one instance noticed in these pages it is nearly two and a fifth times the diameter. The areas, therefore, of the cylinders vary from 1 to 3, to 1 to 6. The steam enters the small cylinder at a high tension, and having pressed the first piston to the end of the stroke, escapes into a jacket, or into a low-pressure casing, whence it then enters the larger cylinder, and completes its work there by expansion, being finally condensed in the usual manner.

In some of these engines (Parker and Weston's, and Tangye's) the two steam-cylinders and pump-cylinders are in a line, the low-pressure cylinder being between the high-pressure and pump-cylinders. The steam-cylinders have a piston-rod common to both, and attached to the pump-piston, or ram, as the case may be, in the same line. In Mr. Davey's 'Differential' and the 'Self-Governing' engines there are two piston-rods to the low-pressure piston, which are united to a high-pressure piston and pump piston-rods by a crosshead, the high-pressure cylinder in these machines being next to the pump. Each steam-cylinder has a steam-chest and a set of valves receiving independent motion from their respective steam-cylinders. The peculiarity of the 'Self-Governing' engine consists in the arrangement of the valve-gear, which is Messrs. Cope and Maxwell's patent, and, like Mr. Henry Davey's, it acts upon the principle of controlling the rate of movement of the engine by the flow of a liquid from one end to the other of a cataract-cylinder.

We purpose to notice in detail the 'Differential' valve-gear; the Self-Governing' valve-gear has already been described. In Cherry's pump the low-pressure cylinder surrounds the high-pressure, one ordinary flat-faced slide-valve and one steam-chest only being used for both, while the space occupied is no more than that of an ordinary single-cylinder engine. The low-pressure cylinder has two piston-rods in a horizontal line with the high-pressure piston-rod, and the three rods are attached to a crosshead, which also carries the pump-rod.

Sturgeon's pump consists of a low-pressure cylinder in a straight line with the pump, combined with a pair of high-pressure cylinders, one on each side of the low-pressure. The three piston-rods are attached to a crosshead, the middle rod being continued through to the pump-piston. The valve-chest is placed upon the top of the low-pressure cylinder.

The pumps of Messrs. Field and Cotton and Mr. Walker are very peculiar, and are novelties in the way of compound engines. The former was exhibited in model form at the Vienna Exhibition in 1873. It consists of a high-pressure cylinder placed within a low-pressure, the former moving forwards and backwards upon a fixed piston attached to a piston-rod. The high-pressure cylinder, being fitted with piston heads at each end, forms a piston for the low-pressure cylinder, the annular space between the piston-heads forming a steam-chamber to which the live steam has constant access. The whole arrangement is so unique and so peculiar that an intelligible idea of it can only be conveyed by a detailed description and drawings.

Walker's pump consists of a steam-cylinder, with an annular partition fixed at the centre of its length. A barrel of considerable length, with piston-heads at each end, is fitted into the cylinder; the barrel part works steam-tight in the latter. The valve-chest is fixed upon the cylinder. There is only one piston-rod. This machine too requires drawings and a detailed description to make any account of it intelligible.

In pumps such as Cherry's and Sturgeon's, a great saving in length is effected compared with those which have the cylinders end to end. This is an important feature in mining pumps for use underground. It also effects a considerable saving in foundations. The dimensions of Sturgeon's pump of two high-pressure cylinders, 11 inches diameter, and one low-pressure of 26 inches diameter, with a 2-feet stroke, working two 7-inch double-acting rams, are: length over all, 13 feet 8 inches; width, 5 feet 2 inches; height to the top of the air-vessel, 8 feet. A pump of this size is capable of throwing 10,000 gallons per hour 600 feet. The steam-passages between the high and low-pressure cylinders can be made very short, and the loss in pressure owing to the steam having to fill the long pipe and extra valve-box in the end-to-end cylinder arrangement is avoided. In these pumps a single valve and valve-gear serve the purpose of two distinct and separate valves and valve-gears on the other system.

Tangye's Compound Engine.

Diameter of high-pressure cyl. in.	8	8	10	10	12	12	12	14	14	14	16	16	16	18	18	18	21	21	24	24	30	30
Diameter of low-pressure cyl. in.	14	14	18	18	21	21	21	24	24	24	28	28	28	32	32	32	36	36	42	42	52	52
Diameter of water-cylinder . in.	4	6	5	8	6	8	10	7	10	12	8	12	14	8	12	14	10	14	10	14	12	14
Length of stroke in.	24	24	24	24	24	24	24	36	36	36	36	36	36	48	48	48	48	48	48	48	48	48
Delivery per hour, approx. galls.	3,900	8,800	6,100	15,650	8,800	15,650	24,450	12,000	24,450	35,225	15,650	35,225	47,950	13,650	35,225	47,950	24,450	47,950	24,450	47,050	35,225	47,950
Diameter of suction and delivery in.	3	4	3½	6	4	6	8	5	8	9	6	9	10	6	9	10	8	10	8	10	9	10
Diameter of high-pressure steam-inlet . . in.	1¼	1¼	1½	1½	2¼	2¼	2¼	2¼	2¼	2¼	2½	2½	2½	3	3	3	3½	3½	4	4	5½	5½
Diameter of low-pressure steam-exhaust . in.	1½	1½	1¾	1¾	2½	2½	2½	2½	2½	2½	3	3	3	3½	3½	3½	4	4	5	5	6½	6½
Height water can be raised with 40 lbs. steam-pressure non-condensing . . feet	360	160	360	140	360	202	130	360	175	122	360	160	118	456	202	149	397	202	518	264	562	413
Height water can be raised with Holman's condenser . . feet	480	213	480	187	480	269	173	480	234	162	480	213	154	603	269	198	528	269	691	352	750	550
Height water can be raised with air-pump condenser . . feet	600	267	600	335	600	337	216	600	203	203	600	267	191	750	337	248	660	337	864	440	937	689

Walker's Compound Engine.

Diameter of steam-cylinder . . in.	8	8	8	8	8	8	10	10	10	10	10	12	12	12	12	12
Diameter of water-cylinder . . in.	3	4	5	6	7	8	6	7	8	9	10	7	8	9	10	12
Gallons per hour .	1,830	3,250	5,070	7,330	9,750	13,000	7,330	9,750	13,000	16,519	20,000	9,750	13,000	16,519	20,000	30,000
Price . . £	£40	£45	£50	£55	£60	£70	£65	£70	£80	£95	£110	£85	£95	£115	£130	£150

Sizes.

These pumps can be made of any size, and any combinations can be made between the steam and pump-cylinders. The smallest and largest sizes advertised by the makers are 6 × 14 inches to 35 × 60 inches for the high and low-pressure steam-cylinders respectively, and from 3 to 14 inches for the pump-cylinders. The stroke ranges from 2 to 6 feet.

The tables opposite, of some of the sizes of the compound pumping engines made by Messrs. Tangye Brothers, and Holman, and Clayton, Howlett, and Venables (Walker's Pump), afford much useful information relative to the sizes and capacities of pumps of this class:

Capabilities.

By referring to the tables of the sizes, &c., of Tangye's and Walker's compound engines the reader will observe the quantities of water that may be raised by engines of given sizes. The largest sizes, we perceive, with the best form of condenser, will raise to nearly a height of 700 feet about 48,000 gallons of water in an hour. The 'Differential' compound steam-pump of Mr. Henry Davey, of Leeds, has been made for use underground in sizes as large as 35 inches high-pressure cylinder, and 60 inches low-pressure cylinder, with a 6-feet stroke, and capable of delivering more than 40,000 gallons per hour about 900 feet high.

The following account of the work done by engines of the Compound 'Differential' type, applied underground in collieries, is extracted from papers courteously supplied by Mr. Henry Davey, the patentee:

Cylinders		Stroke	Particulars of work
High pressure	Low pressure		
in.	in.	in.	
33	50	72	To raise 37,200 gallons per hour 1,200 feet.
22	54	72	A pair, each to raise 30,000 gallons per hour 920 feet, and to give power to a hydraulic engine which lifts 30,000 gallons per hour 64 feet.
25	50	60	30,000 gallons per hour 450 feet.
28	50	60	24,000 gallons per hour 1,100 feet.
33	54	72	3, each to raise 60,000 gallons per hour 345 feet.
35	60	72	A pair, each to raise 42,000 gallons per hour 910 feet.
33	54	72	To raise 72,000 gallons per hour 300 feet.

These pumps may be supplied with steam from the surface, or from a boiler near the engine underground, but it is better to get the steam from the surface; this is generally done. It is possible to use compressed air, but its useful effect is much less than that of steam.

The Chesterfield and Boythorpe Colliery Company have one of Tangye's Double-ram Compound Engines at work 400 feet below the surface, the steam being taken down to it at a pressure of 45 lbs. per square inch. The engine has 21-inch and 36-inch steam-cylinders, 12-inch rams, and a 4-feet stroke. It can be worked without any difficulty at 28 strokes per minute (equal to 224 feet piston speed). The vacuum in the condenser is from 11½ to 13 lbs.

The illustration opposite, fig. 89, represents Mr. Davey's system of draining collieries and other mines, as carried out on a large scale in a colliery 1,200 feet deep.

At a point in the pit 900 feet from the surface are placed a pair of compound 'Differential' engines, pumps, and a separate condenser. The engines have cylinders 35 inches and 60 inches in diameter, and a 6-feet stroke. The pump-rams, which are of gun-metal, are 12½ inches in diameter. At the bottom of the pit, 300 feet below the main-engines, are placed a pair of hydraulic pumping engines, which are to lift 60,000 gallons per hour to the main-engines. The latter force the water to the surface and supply power through the

Fig. 89.

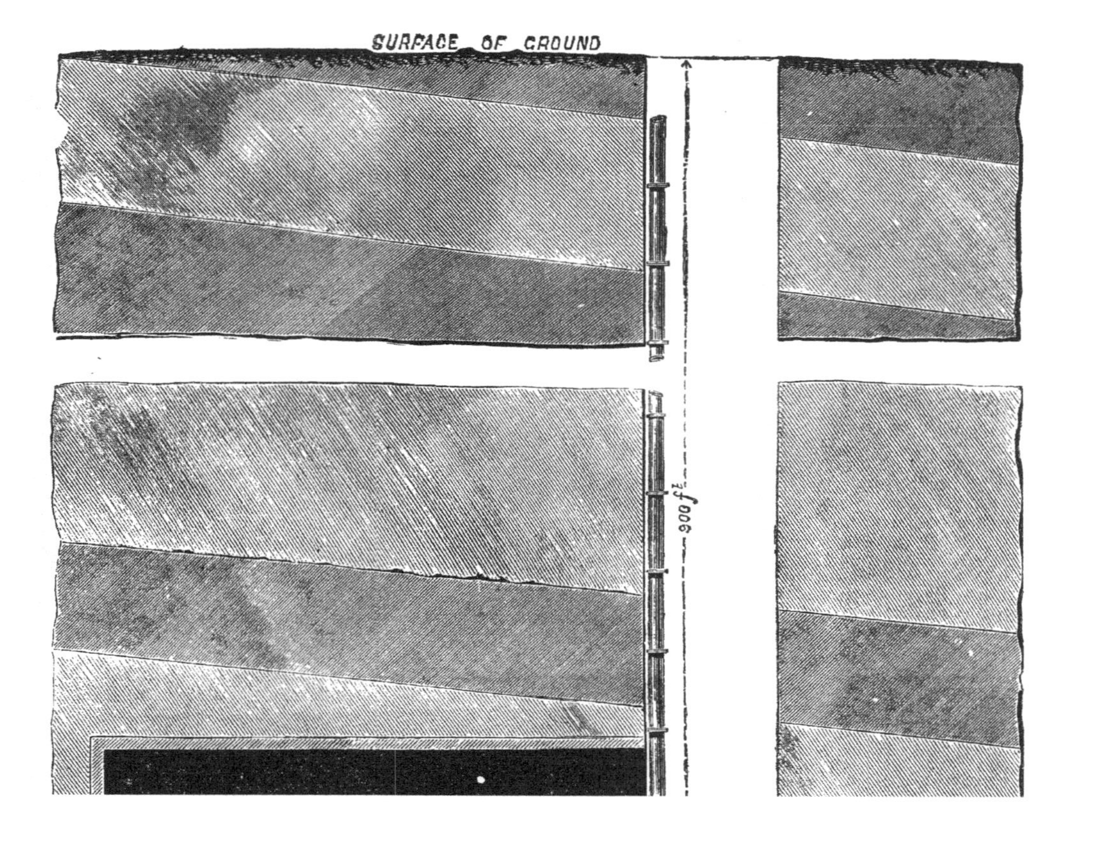

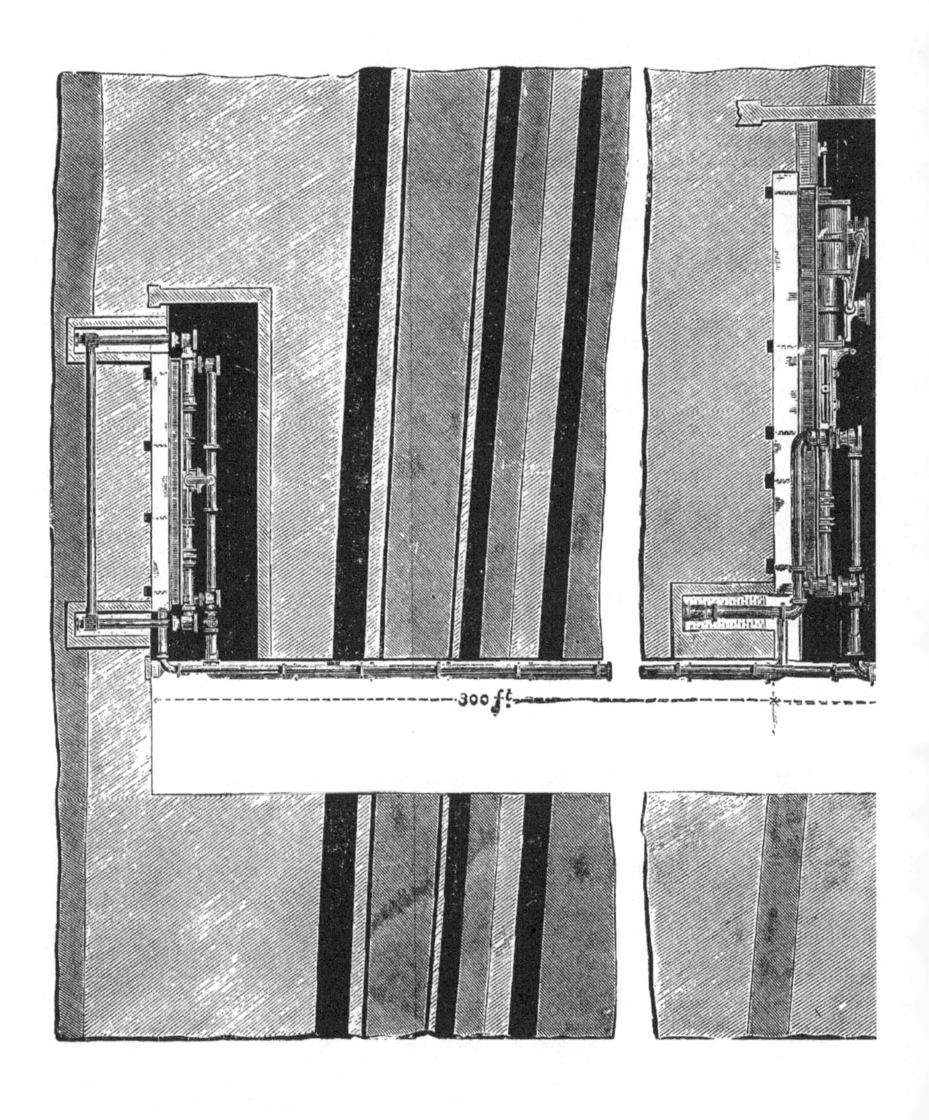

300 ft.

column to the hydraulic engines. By this system the main-engines are kept out of danger of flooding. The hydraulic engines will work under water, and can be actuated from the main-engine room. As further security, they could be placed in a water-tight chamber, accessible from the main engine room, through a water-tight staple. By such means the hydraulic engines could be under repair even if the water rose to the main-engines, 300 feet up the shaft. The principle of its action is that of employing water at a given head to raise a larger quantity against a less head.

'Dip workings are drained by means of hydraulic engines applied on the same principle . . . and a great number are working very successfully in various collieries. The useful effect is about 80 per cent. of the power applied.'

We have already stated that these engines are generally supplied with steam from the surface. The reasons for preferring to do this to giving them steam from boilers alongside are stated at page 143. Of course loss from condensation in the latter case is avoided, but the disadvantages more than counterbalance this gain.

'These engines are not so suitable for metalliferous mines as they are for collieries. In many collieries one shaft is made to serve the double purpose of pumping and winding, and there is not sufficient room for the pump-rods and the winding apparatus, and the underground engine in such a case becomes almost a necessity. The extra amount of steam used by the condensation in the pipes is more than fully compensated by the advantages of having no rods in the pit to interfere with the winding arrangements, and of having a very much smaller amount of machinery to do the work. Underground engines and pumps are very much smaller than those on the surface, because the absence of rods enables them to be driven at much higher speeds.'

It must be borne in mind that it is not practicable to sink pits with engines of this description without additional pumping apparatus, their sphere being limited to collieries where the required depth has already been attained. We say *collieries* advisedly, because the cost of fuel prevents the use of large underground engines in metalliferous mines (see pp. 136, 164, and 165).

Cost.

A very important consideration in the use of steam-engines is their first cost, but the weight to be attached to this feature of the engine depends very much upon circumstances. In collieries where large quantities of unsaleable slack are at command, or other low-quality coal which costs only a few pence per ton to raise, the first cost of an engine is thought to be of more consequence than its working results; but in a metalliferous district, such as Cornwall, with the cost of coal at 15*s.* per ton, the first cost is of secondary importance, and the supreme effort is how to get the water out of the mine with the smallest consumption of coal. The heaviest item of dead cost in a metalliferous mine is generally that of pumping. It is evident under such circumstances that an economical engine compared with a less economical one would ultimately pay the excess of first cost, the length of time depending upon the relative economical values of the engines.

Another phase of the question has to be considered. It sometimes happens that the convenience which attends the using of an engine of a particular kind will compensate for its inferior economy. This is especially the case with some forms of underground engines which can be used in places and under circumstances where it would not be easy nor even perhaps possible to use any other kind of engine. In this case we have not only the greater convenience but the lesser first cost of the engine. These are features of the question which the capitalist must consider in the choice of machinery for the unwatering of mines.

In order to give the relative cost of the different engines it will be necessary to arrive at the cost of engines for the performance of a given quantity of work. We will assume then that the quantity of water to be raised is 30,000 gallons per hour, and the height 1,800 feet, which is about the depth of the deepest mine in Cornwall (from the adit) and the deeper coal mines.

A Cornish engine of the following dimensions would be required for the performance of the quantity of work stated—

	tons	
95-inch cylinder, 11-feet stroke in the cylinder, and 10-feet in the shaft, working at 8½ strokes per minute .	133	£2,925
Erecting ditto and boilers		350
4–14-ton boilers and outfits.	64	1,365
Engine and boiler houses		800
	197	£5,440
Pitwork.		
Pitwork 13½ inches, 6 plunger-lifts, each 50 fms. .	195	£2,050
Fixing ditto		350
Rods averaging 17 inches square.	50	310
Rod-plates	38	830
2 balance-beams	12	220
	295	£3,760
Total tons	492	£9,200

The following estimate of the cost of the 'Differential' Underground Compound Engine and pumps and hydraulic engine, for the same work, is furnished by Mr. Davey:

	tons	£
Engine and pumps	50	1,500
Hydraulic engine	12½	500
Steam and water pipes	160	1,600
Boilers	*40	1,200
Buildings, excavation, &c.	—	1,000
	262½	£5,800

The estimate of the Underground engine only applies when the pit is *already sunk*. Till then, the engine must be used at surface, as described at page 164, and be afterwards taken underground and the pumping apparatus thrown aside and new substituted. If the first cost of the engine, and the cost of removing it and the pumpwork, and the cost of new pump-

* This is only about two-thirds the weight of the boilers for the Cornish engine, which is a more economical machine. The explanation may be that Mr. Davey's boilers are multitubular, the heating capacity of which is much greater than that of Cornish boilers, which are designed specially for slow combustion, and are consequently large in proportion to the engine.

work, be added, the total would probably reach that of the Cornish engine. Moreover, the working economy of the Underground engine not being nearly so good as that of the Cornish, it is evident it will not bear a comparison with the latter for sinking operations. The circumstances which necessitate the adoption of underground engines in collieries already down to the required depth have been noticed at pp. 164 and 165.

The following is the cost of Davey's Surface Engine at the South Durham Colliery: *

'Cylinders, 45 and 72-in. 10-feet stroke. Actual H.P. 550; 10 strokes per minute; plungers 20 in. dia.: column 18 in. Work, 156,000 gallons per hour, raised 700 feet. Weight of quadrants 25 tons, pumps 60 tons, column 100 tons, and engines 66 tons. The cost of the engine and pumps was £6,000, and foundation and engine-house £600—total £6,600.

'A Cornish engine for the same work: Cylinder, 100-in.; stroke 10 feet; strokes per minute 10; plungers 28¼ in. dia.; column 26 in. Engine 160 tons; pumps 220 tons. The cost of the engine and pumps would be £6,700, and foundation and engine-house £1,000—total £7,700.'

Duty.

It is much to be regretted that so little information concerning the performance of these machines in practice is obtainable. The author has experienced great difficulty in obtaining reliable data on this and the preceding class of machines. For some reason which he cannot pretend to comprehend the managers and engineers of mines in which these machines are at work—persons who of all others ought to be interested and concerned in the matter—object, as a rule, to give any account of their engines such as would be required for comparing their performances with other machines, and so promote the adoption of the most efficient machines. There appears amongst the mining community generally a lack of interest in this important question, one of the most important

* *Minutes of Proceedings of Inst. Civil Engineers*, vol. liii., Session 1877-8, part 3, 'Davey on Pumping Engines,' p. 9.

connected with pumping machinery, and as long as persons who have engines at work upon their premises under a variety of circumstances and for a lengthy period, remain indifferent to the manner in which those engines get through their work and the return they make for their purchase and cost of maintenance, so long will there be conflicting opinions, unsettled views, and undetermined judgments with regard to their comparative worth. We have now little or nothing more to aid us in setting forth the duties of these engines than the remarks which manufacturers have made concerning their own machinery, desultory discussions of engineers at meetings of scientific societies, and what may be assumed from the construction and principle of the machines.

It is sometimes very difficult to arrive at a knowledge of the duty of these engines, owing to their being worked either by old worn-out boilers, or by boilers which also supply steam to other engines. This we admit, but there are so many instances where the test can be made, and others where a comparatively trifling expense would provide the engine with an independent supply of steam, that it would be very easy to make tests of all the types of machines in use.

The only true test of economy is the expense not only of fuel and stores, but of accidents, and all repairs for a lengthy period. The performance of a machine for a short period cannot be deemed a satisfactory or reliable test of its efficiency, as it might for a short time do a high ratio of duty that would ultimately break it down, or at least so injure it as to require expensive repairs. A few hours' or even days' experiment is not a sufficient trial of the capabilities of an engine, as one that would do a low ratio of duty might really be a more economical machine than one that would do a higher duty, if in the one case we have a steady and continuous performance of work, and in the other frequent breakages or derangements, involving loss of time and money to repair. That engine is the most economical that will do the most work with the least amount of wear and tear.

It will be appropriate to consider, before proceeding further, the conditions necessary to economy, and then to observe to what extent those conditions exist in the class of engines under consideration.

1. The performance of work with a small expenditure of fuel and lubricants is undoubtedly the first and chief step towards economy in working results. The expansive use of steam is the principal means of obtaining economy in this direction, and all other conditions being equal, that engine is the most economical that affords the greatest facility for using high degrees of expansion.

2. The next condition necessary to economy is durability. A temporary high ratio of duty will not make an engine an economical one in the broad sense we have indicated; there must be a long-continued performance of a fair duty, and the engine must be made of such good materials and be so constructed as to remain in a good workable state for a considerable period.

3. The third condition is that the cost of manufacture shall bear a favourable relation to the effective power of the engine, its working results, and its durability.

Recapitulating, we have as the three conditions necessary to economy:

A high ratio of duty.
Durability.
Small first cost.

The engine, then, that will do the highest duty, or, what comes to the same thing, provides for the greatest expansive use of steam, that costs least to keep in working order, that will remain in a good workable state for the longest period, and that costs least to manufacture, is the most economical. But it happens that all these qualities are not found in one machine, and in order to determine the relative economical values of machines it will be necessary to discover to what extent they severally combine these conditions.

The compound engine* will admit of being worked under

* It must be understood we do not here refer to the compound engine used underground.

degrees of expansion that would be quite impracticable in a single-cylinder engine, without an enormous moving mass and a very high initial pressure and speed. A high pressure of steam, high degrees of expansion, and considerable piston speeds (which are conducive to economy) are attainable without either a high maximum piston speed, a heavy initial strain, or a very large moving mass. It is claimed for the compound engine that at an expansion of ten times it produces less variation than a single-cylinder engine cutting off at 40 per cent. of the stroke, and the latter requires a greater momentum in the moving mass.

The following table shows the value of various degrees of expansion:

Initial pressure per sq. in. in lbs.	Ratio of expansion	Units of work	Average pressure in lbs.	Proportion of stroke at which steam is cut off	Efficiency of engine throughout stroke being 1,000
100	0	63000	100	$\frac{2}{3}$	1·405
100	1·25	77017·5	97·8	$\frac{1}{2}$	1·693
100	1·66	94958·64	90·8	$\frac{1}{3}$	2·099
100	2	106596	84·6	$\frac{1}{4}$	2·386
100	3	132111	69·9	$\frac{1}{5}$	2·609
100	4	150192	59·6	$\frac{1}{6}$	2·792
100	5	164115	52·1	$\frac{1}{7}$	2·942
100	8	193536	38·4	—	—
100	10	207900	33	—	—

The variation in force between the commencement and the end of the stroke is as $2\frac{1}{2}$ to 1; whereas in the single-cylinder, with the same degree of expansion, it would be as 6·3 to 1; that is to say, the variation in the two pressures is nearly three times as great in the one as in the other. The importance of thus reducing the strains on the machinery cannot be overlooked. The engine may be made lighter with greater security against breakages, the foundations are cheaper, and the speed of the engine is more uniform.

The accompanying table* contains a comparison of the initial, terminal, and average pressures, maximum piston. speeds, and proportions of weight of mass to the water-load:

* Inst. Mechanical Engineers, *Proceedings*, 1872, p. 125.

	Type of engine	Pressures: Initial absolute	Terminal absolute	Average	Initial ÷ terminal	Average ÷ terminal	Forces: Initial ÷ Mean	Piston velocities: Maximum velocities per minute	Mean water speed per minute	Mass ÷ water load
		lbs.	lbs.	lbs.	lbs.	lbs.		ft.	ft.	
a	Cornish	45	10	19	4·5	1·9	2·26	500	80	3 about
b	Ditto	31	10	16·1	3·1	1·6	1·8	600	—	1·7
c	Ditto	25	9	12·2	2·77	1·35	1·72	600	—	1·7
d	Compd. differential	80	10	24·1	10·66	2·4	1·37	220	150	2
e	Ditto	43·7	7	12·75	6·24	1·8	1·4	228	168	0·66
f	Ditto	85	11·3	20·25	7·5	1·78	1·3	—	—	1·1
g	Ditto	25	8·5	14	—	1·6	—	210	144	1·2
h	Single cylinder differential	30	17	19·6	1·75	1·15	1·4	190	130	1·1
i	Corliss	45	6	16·97	7·5	2·8	2·53	—	—	—
j	Cornish	—	—	13 abt.	—	—	—	600	100	1·7
k	Ditto	—	—	—	—	—	—	570	112	1·7

The foregoing shows how easily high-pressure steam and great degrees of expansion can be used in compound direct-acting engines without a high maximum piston speed or a large moving mass, both of which are necessary in the single-cylinder engine. The highest initial pressure in the Cornish engine is in the case of an engine (*a*) having a mass equal to three times the water-load. The initial pressure is not more than 45 lbs. per square inch. In the compound engine (*d*) the initial pressure is 80 lbs. and the mass is only twice the water-load. The terminal pressure is the same in both cases. From what has been stated of the facility with which expansion can be carried out in a compound engine, and the high ratios of the expansion, we shall be prepared to learn that this engine is capable of performing a high duty.

Mr. Bramwell, in a paper on the 'Economy of Fuel in Steam Navigation,' supplied a table of nineteen engines which had an average consumption in long sea voyages of 2·11 lbs. of coal per indicated horse-power per hour, while five of the engines consumed less than 2 lbs. of coal. The absolute boiler pressure ranged from 63 to 75 lbs. per square inch, and appears to have been admitted to the high-pressure pistons at nearly the same pressure. Marine engines on the newest construction do not consume, it seems, more than from about $1\frac{3}{4}$ to 2 lbs. per indicated horse-power per hour. In the case of marine engines we have the compound engine placed under the most favourable circumstances for performing a high duty, and their performances may be taken as the utmost useful effect that can be got from compound engines. The boilers and the engines are in proximity, and the latter are brought direct to the work to be done, and can be run at a very great speed. In the types of compound 'differential' pumping engines of Mr. Henry Davey, which are worked at the surface, and supplied with steam from boilers alongside, a high ratio of duty can also be performed. As an example, an engine of the Croydon type, working at a colliery with boilers not roofed over and the steam jacketing far from perfect, consumed 22 lbs. of water per indicated horse-power per hour, equal to about 2·3 lbs. of coal, or about 70,000,000 duty, the steam-pressure being 80 lbs. (absolute) per square inch.

As the type of engine now under consideration is applied direct to the water by placing it underground, we have to consider the duty it will under such a circumstance perform. When an engine is placed down a deep pit one of two alternatives must be adopted. The steam must be conveyed to the engine from the surface, or it must be supplied from a boiler placed underground near the engine. The former method is most generally adopted. It is awkward to have a boiler underground; one reason is that it causes such a heat in the workings, and makes it so intolerably hot to the persons in charge of the engines, and in fiery collieries it is dangerous to have boilers in or near the workings.

Assuming the marine compound engine to perform a duty of 2 lbs. of coal per indicated horse-power per hour, and bearing in mind that much of that high duty is due to the high pressure of the steam, the very great rapidity (say 400 feet per minute) the piston is driven, and the proximity of the boiler to the engine, we have, in the case of the compound engine used underground for pumping, and supplied with steam from the surface, to make three allowances:

1st, for a diminished piston speed, say about one-half of that of the marine engine, owing to the speed of the steam and water-pistons being identical;

2nd, for loss of steam in condensation. It appears from actual results that the loss of steam in condensation under the most favourable circumstances when taken down a deep pit amounts to about ½ lb. of coal per indicated horse-power per hour. Take a case in point. 'A pair of "differential" engines with 20-inch cylinders are supplied with steam from the surface in 7½-inch pipes, clothed with non-conducting cement, of a total length of 1,100 feet. The loss of steam from condensation was found to amount to 8 cubic feet of water per hour when the engines are standing, and 12 cubic feet when working at the ordinary speed of 10 double strokes per minute, with 45 lbs. of steam. . . The loss of coal from this condensation in the steam-pipe amounts to 2,160 lbs. in 24 hours if the engines are standing half the time, or 2,592 lbs. if working continuously.' This loss is equivalent to about ½ lb. of coal per indicated horse-power.

At Tynewydd Colliery, where the inundation occurred in April 1877, the loss in conveying steam 130 yards through a 2½-inch pipe was 8 lbs. per square inch, the steam being supplied at 110 and used at 102 lbs. per square inch.

3rd, for the smaller amount of expansion available owing to the loss of pressure by condensation and friction. In an underground compound engine it is necessary to carry full steam the whole length of the stroke in the high-pressure and to expand it in the low-pressure cylinder. Without a very high initial pressure of steam the moving mass is not sufficient to permit of any expansion in the first cylinder. This practically means that an engine of this kind, taking the areas of the high and low-pressure cylinders as 1 to 6, is only capable of an expansion equivalent to the steam being cut off at one-fifth of the stroke in a single-cylinder engine. The accompanying indicator diagram, fig. 90, shows the expansion in a 22-inch and 54-inch engine. Taking into consideration these serious drawbacks as affecting the working economy, it is doubtful if underground compound engines will yield a better result under ordinary circumstances than 4 lbs. of the best steam-coal per indicated horse-power.

FIG. 90.

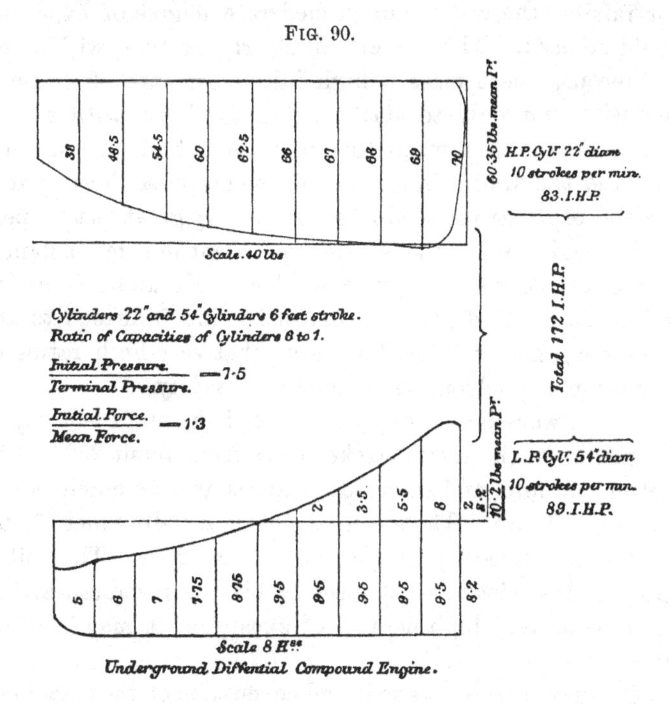

Underground Differential Compound Engine.

Let us now examine to what extent the expansion of steam is practicable in a single-cylinder pumping engine, with particular reference to the Cornish engine.

The conditions necessary to expansion are:

1. *A heavy moving mass* (to absorb the excess of pressure at the beginning of the stroke, like a fly-wheel, and to give it out towards the end of the stroke to compensate for the diminishing pressure of the steam).

2. *A great initial pressure of steam* (to start the moving mass). The moving mass in the Cornish engine consists of the large main-beam, heavy pump-rods, plungers, &c., and varies according to the depth of the mine, from twice to five or six times the water-load, but it is usually perhaps not more than three times. In one extreme instance it amounted to about 11½ times the water-load. The ratio of expansion is in proportion to the excess of the moving mass and the pressure of the steam. If the moving mass were only that sufficient for raising the water, only a moderate degree of expansion could be used. The reader will clearly see that, with a very ponderous, inert mass a high initial pressure of steam is requisite, not only to start it, but to do so with a sufficient impetus that when the steam is cut off the momentum and the diminishing force of the steam shall carry it to the end of its stroke. This leads to a very great piston speed at the beginning of the stroke, equivalent in some instances to 800 feet per minute, and a sudden strain and concussion which at high degrees of expansion are injurious to the machinery, and it is for this reason that very high ratios of expansion can seldom be adopted with safety.

When working at its greatest speed the mean velocity of the piston on the steam stroke varies from about 250 to 350 feet per minute, and in some instances it is as much as 600 feet per minute. The steam stroke is usually much faster than that caused by the descent of the rods. The initial speed of the piston is at least double the terminal, and in the case of very high degrees of expansion it may be three times.

Owing to the great strain and concussion of the machinery at the beginning of the stroke, caused by the use of very high-pressure steam, steam of a moderate pressure, compared with that which has been adopted in compound engines, is used in the single-cylinder engine. The highest pressure in a Cornish engine is not above 45 or 50 lbs. per square inch, which is about two-thirds of that used in compound engines.

The following reference to the foregoing table of pressures,

&c., shows the greater facility which the compound engine offers for the expansion of steam :

	Cornish engine			Compound 'differential' surface engine	
	1st ex	2nd ex.	3rd ex.	1st ex.	2nd ex.
Initial steam-pressure (absolute) lbs.	45	31	—	80	43·7
Terminal ditto . . . . lbs.	10	10	—	10	7
Maximum piston-speed. . feet	500	600	570	220	228
Mean water-speed. . . feet	80	—	112	150	168
Proportion of mass to water-load .	3 to 1	1·7 to 1	1·7 to 1	2 to 1	·66 to 1

It must be conceded, notwithstanding the above examples, which do not fairly represent its merits, that the Cornish engine admits of the greatest economy in expanding steam. The momentum which can be stored up in the moving mass is sufficient to keep the engine in motion at a lower pressure of steam than is possible in any engine not having such a preponderating inert weight. With a mean water-load of 12 lbs. per square-inch the steam may be expanded down to from 5 to 3 lbs. Taking this, and the loss of pressure in expanding in two cylinders, into account, it is evident there must be a much higher ratio of expansion in a compound than in a Cornish engine to produce the same useful effect, and steam of a higher pressure is condensed.

'It was about the year 1785 that Watt introduced the principle of the expansion of steam into Cornwall, but the steam was not raised higher than before, and the piston made a considerable part of the stroke, therefore, before the steam-valve was closed.' * Great strides were made in this direction after Mr. Watt's death, and high degrees of expansion led to high duties being performed. The highest ratio of expansion has been, we believe, ten times, but it is very rare for such a great expansion to be attempted. The celebrated Taylor's engine, at the United Mines, Gwennap, worked at a cut-off of one-tenth of the stroke, and performed the very high

* Mr. Taylor's *Records of Mining*.

duty of 107,500,000 lbs. raised 1 foot high by the combustion of a bushel of coal (94 lbs.)* Another Cornish engine—Austen's, at Fowey Consols Mine—as the result of a 24 hours' experiment, did a duty of 125,000,000 lbs., and 90,000,000 and nearly 98,000,000 in two months for the same quantity of coal. In 1834 the average duty of the best engines for the year was about 91,000,000 lbs. In the following year Austen's engine did an average duty of 91,672,210 lbs.† In September 1836 the Wheal Darlington engine performed an average duty of 95,000,000 lbs. for the month. The system of reporting the engines having fallen in a great measure into disuse, and the same care not now being taken to produce great results, and many of the engines being very old, the shafts deeper, and a greater number of them inclined, the average duty is now only about 50,000,000, and the greatest cut-off is about one-fifth of the stroke. Lean's 'Reporter' for Feb. 1871 shows an average duty for the month of nearly 56,000,000, the highest being 79,400,000. The highest duty ever recorded, 125,000,000, is equal to a consumption of 1·49 and about 1·12 lbs. of coal per effective and indicated horse-power per hour respectively; 107,500,000 is equal to 1·73 and 1·29 lbs., and 90,000,000, which we believe the engine, under favourable circumstances, is able to do continuously, is equal to 2·06 and 1·55 lbs per effective and indicated horse-power per hour respectively. Having a due regard, however, to the durability of the engine and the pump-work, it is a wiser economy to work at a moderate grade of expansion, and to realise, say, a duty of 70,000,000, for 112 lbs. of coal, equivalent to 3·16 and 2·40 lbs. of coal per effective and indicated horse-power per hour respectively. ‡

* Previous to July, 1856, the duty was computed per bushel (94 lbs.) of coal, but since that date it has been computed per cwt. (112 lbs.)

† This engine is now in the north of Eugland, and is believed to be still at work.

‡ 'From an examination of several indicator diagrams of pumping engines, I have come to the conclusion that the difference between the indicated horse-power and the effective varies from 25 to 50 per cent.; therefore, for the sake of comparing the duty performed by different

Comparing the duty of the Compound Engine used underground as far as we can ascertain it with the Cornish Engine, as a good example of an economical surface-pumping engine, we have approximately—

Compound engine	.	4 lbs. of coal per indicated horse-power per hour.
Cornish engine .	.	2·40 ,, ,, ,,

Compound rotative engines are reputed to do a very high duty. Engines of this kind erected at the Lambeth and Chelsea Waterworks are said to have attained the very high duty of 111,350,000, or equal to about 2 lbs. per horse-power per hour of effective, or 1·61 lbs. of indicated duty. They seem to be admirably adapted for water-works, but they are not considered to be suitable for deep mine drainage. When working at a low speed a slow motion is given to the pumps compared with the brisk stroke of the Cornish and some other engines working at the same number of strokes per minute. A great slip thereby occurs through the valves, and the quantity of water actually lifted is much less in proportion to the theoretical quantity than in other machines.

Durability.

It scarcely need be observed that the extent to which a machine is durable is a very important consideration, and one which ought to be carefully taken into account in a comparison of the ultimate values of engines. An engine which may perform a very satisfactory duty in a given period of time may yet in its construction contain a weakness, inherent perhaps to the class of engines to which it may

engines, the indicated as well as the effective horse-power should be reported. When comparing the duty performed by pumping engines, and those employed for marine or other purposes, it must be borne in mind that the first is effective and the latter indicated horse-power. Thus, corrected for friction, the consumption of coal per indicated horse-power per hour, when in 1844 the average duty was 68 millions, did not probably exceed 2¼ lbs., a result not surpassed by the average working of the best compound engines of the present day consuming best steam-coals.' (Mr. W. Husband on Pumping Machinery, vide 'Proceedings Mining Institute of Cornwall,' pp. 173–4.)

belong, and which from the principle of its action it may be impossible to avoid. In mining it is often the case that time is money: night and day there is much 'dead' cost going on whether mineral be produced or not. In Cornish and other metalliferous mines the bottom workings are often the most valuable, and they can be operated upon only when the pumping engine is at work keeping them drained. A derangement, then, of the pumping apparatus leads to the water filling the bottom of the mine and suspending all work in that part, and the evil will be in proportion to the length and frequency of the stoppages of the engine. A protracted stoppage in a heavily-watered mine often leads to serious results; not unseldom is it the cause of a train of other misfortunes, besides the flooding of the mineral veins, such as the running together of ill-secured levels, when the stuff is of a loose nature, the choking-up of adits, &c. It will thus be seen how important it is that the pumping arrangement be of a durable character, and little liable to derangement, and not requiring frequent stoppages for repairs, and renewals of worn parts.

The durability of no engine has been so thoroughly tested as that of the Cornish, the durable nature of which is so well known. A few years ago, when most of the Cornish mines were working, engines could be seen that had been constantly at work for thirty, forty, and even fifty years. The well-known Taylor's engine (85-inch) erected in 1840, was kept almost constantly at work till 1870, and in July of the preceding year it did a duty of 68,750,000 (speed seven strokes per minute, cut-off one-fourth, load 103,200 lbs., or about 15 lbs. per square inch of the piston), thus showing clearly the durable character of the Cornish engine.* The United Mines were abandoned in 1870, or the durable quality of the engine might have been further exemplified. The engine doing such good work at Mellanear is fifty years old, and the famous Austen's engine is believed to be still at work in another part of the kingdom. There can be no doubt that engines with a short stroke and rapid piston motion have a great deal of wear and tear compared with long-stroke engines.

* *Engineer*, vol. xxix., p. 325.

In underground engines, there is more wear and tear in the moving parts and cylinders, particularly when horizontal, than in surface-engines with longer strokes and a slower speed. Underground engines and pumps being much smaller than surface engines and the pumps connected with them must run at a much higher speed to get through the same amount of work, and the stroke being shorter a greater number of strokes is necessary. This leads to the valves of both the engine and the pump being operated a greater number of times, and causes a greater wear and tear in the valves. This is one reason why long-stroke should be preferred to short-stroke engines. Some of the movements of the valves of certain steam-pumps are so delicate that a slight obstruction will suffice to prevent their moving at all. This could not be possible in the Cornish engine, as all the movements of the valve-gear are the result of the action of tappets and levers receiving their motion through plug-rods, actuated by the main-beam, and, consequently, the great power exerted quite overcomes any obstruction. The gear is, moreover, completely under the control of the engineer.

Valves actuated by tappets are more reliable in their action than steam-moved valves, although they may present more wearing surfaces. If in the steam-valves of Cornish engines a breakage should be caused by any means, they can easily be again ground true and thus be kept perfectly steam-tight. Under favourable circumstances the valve-gear of these engines may be in use for many years without needing renewing or even repairing; there are no delicate parts whereas that of many of the steam-pumps is nothing but an assemblage of delicate parts, easily injured or disabled. Steam-pumps being necessarily much smaller than surface engines (being chiefly used for moderate depths and light drainage for experimental purposes), the moving parts are of course correspondingly smaller, and therefore more delicate; in addition to this, the nature of their construction renders them more susceptible to derangement.

Let us regard for a moment the durability of the water-end of these pumps. A piston-pump should only be used when the purity of the water and its freedom from solid impurities can be relied upon, as the wear of the cylinder and piston when mud, coal dross, gravel, sand, or any other solid matter is present is very great, and forms a fatal objection to the permanent use of this kind of pump for such water. The nature of the pumping which has to be performed by the Cornish engine in a progressive mine would very soon disable such a pump. It promotes the durability of the water-part of the machine to employ a ram or plunger instead of a piston; the ram works through the unctuous and soft yet water-tight material of stuffing-boxes, and nowhere touches the cylinder. Other evils are due to water charged with solid matter. The water-ports are liable to be choked with large substances, as pieces of broken rock or coal and wood. The corners around and near the valves are sometimes filled with mud and other solid substances, and the valves prevented from acting properly owing to these substances getting around the seats and between them and the valves. Of course these evils occur more particularly to small pumps.

COMPOUND STEAM-CYLINDER, EXPANSIVE, CONDENSING ENGINES, WORKING DOUBLE-ACTING PUMP.

DAVEY'S PATENT 'DIFFERENTIAL' COMPOUND ENGINE.

The different forms of Mr. Davey's engine have received the appellative 'Differential,' from the very ingenious and valuable self-governing arrangement of the valve-gear for proportioning the supply of steam to the varying pressures of the working load.

'The Differential engine exists in two distinct types, viz., the single-cylinder and the compound engine; the latter, admitting of being worked with very high-pressure steam and high degrees of expansion, is capable of realising the greatest economy of fuel. The chief peculiarity in the invention is the simple manner in which the engine is made perfectly safe in working under all conditions of load, automatically and instantly varying its supply of steam with very minute increase or decrease of resistance, the distribution of steam being such that the pumping is performed without shock, even when the pressures suddenly and greatly vary.'

There are two forms of Single-cylinder engines: the *Differential steam-pump*, and the *single-cylinder Differential pumping engine*, both for use underground. Of Compound-cylinder engines there are three forms: *vertical* and *horizontal engines*, used at the surface for actuating pump-rods, and the *horizontal engine*, applied underground, see fig. 89. The single-cylinder Differential engine and the Differential steam-pump have already been noticed. Before considering the underground compound horizontal engine, let us first explain the nature of that feature in these engines which distinguishes them from other engines—the Differential valve-gear.

'The main principle of the gear is that the valves of the engine have a motion the resultant of two other motions, the first an independent constant motion, and the other the motion of the engine, which may be termed a dependent

FIG. 91.

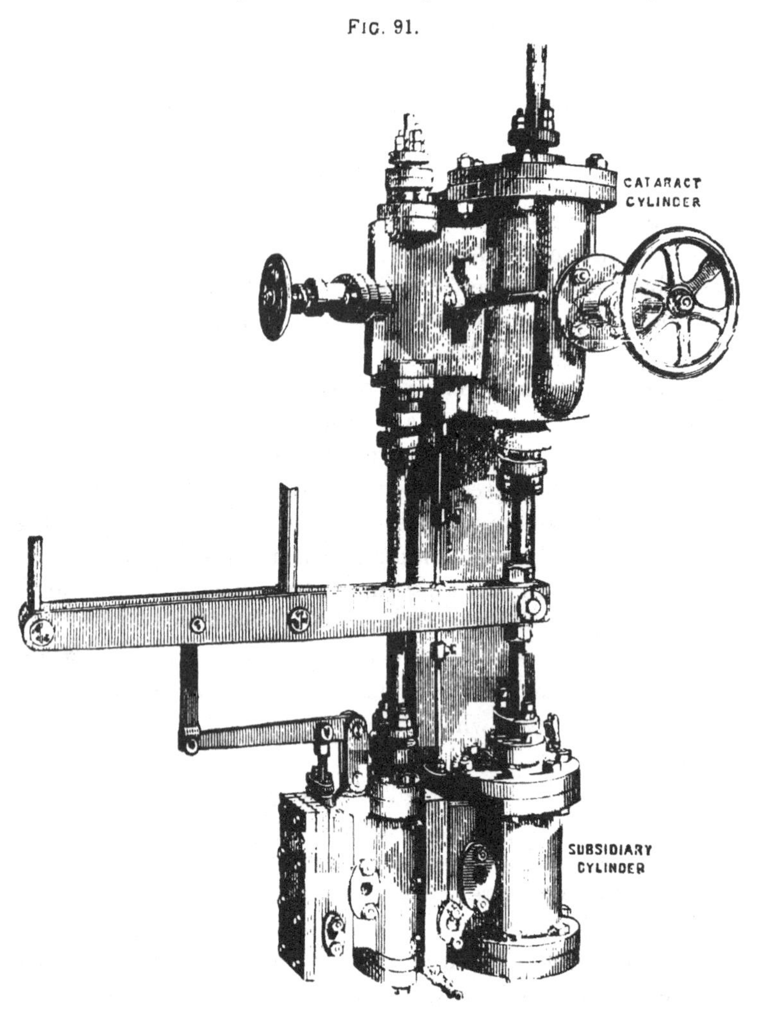

variable motion. Any erratic movement of the engine resulting from a change of load, or other cause, produces a corresponding variation in the distribution of steam through the resultant motion of the valves.'

The principle of the gear is represented by the illustration on the opposite page (fig. 91), which is a perspective view of the gear as applied to engines having drop-valves and to existing Cornish engines. 'It consists of a lever called the main-lever, by means of which motion is given to the valves through a rod. The motion of the engine is given to the outer end of the lever through the rod by means of a lever of the first order, the long end of which is attached to the plug-rod, the other end deriving its motion from the subsidiary cylinder, and being controlled by means of the cataract. The cylinder has a slide-valve, which is worked by means of a tappet arm on the rod of the piston of a secondary cylinder; the motion of the secondary piston is also controlled by a secondary cataract. The slide-valve is, however, free to move with the motion of the hand-lever. It will be seen that there are two hand-wheels and a hand-lever attached to the cataracts. The function of the large wheel is to regulate the speed of the engine during the stroke; the small wheel is for regulating the pause between the strokes, whilst the hand-lever enables the engineman to hand-work the engines. The rocking shaft is employed to give motion to the valves of the engine in the usual way, clearly shown in the engraving.

'The action of the gear may be thus described: Let the engines be "out-doors." The engine end of the main-lever will then be in its highest, and the opposite end in its lowest position, the secondary lever being lifted so as to admit steam to the bottom of the secondary cylinder. The engine will pause until the piston of the secondary cylinder shall have travelled to the end of its stroke, and have lifted the valve of the subsidiary cylinder. The pause will be long or short, according to the regulation of the secondary cataract. The piston of the subsidiary cylinder then having steam on its lower surface will travel upwards, actuating the main-lever with a speed dependent on the adjustment of the cataract. In doing so the steam-valve of the engine will be opened through the medium of the rod, and the rocking-shaft, &c., and will be opened quickly, because the engine end of the lever is for the time being stationary. Steam now being admitted on the engine-piston, it will, after overcoming the inertia of the load, move off at an increasing speed, which is communicated to the engine end of the main-lever. The result is, that as the opposite end of the lever is moving uniformly in the opposite direction at the same time, the motion of the centre is soon reversed in direction, and the steam-valve, which was opened by the motion of the subsidiary-piston, is closed by the differential motion brought into action by the motion of the main-piston acting on the same point.

'It will be seen that the valve of the differential will only be changed by a change in the motion of the engine, because the other component of the differential is a constant, and the time of closing the steam-valve will be earlier or later as the differential is greater or less, and as changes in the motion of the engine are produced by variations of load and steam-pressure, it therefore follows that these variations produce their correction in the distribution of steam.

'If through loss of load the motion of the piston of the engine becomes excessively erratic, steam will, by the action of the gear, be admitted on the reverse side. The full load has often been suddenly thrown off from a differential engine moving at its normal speed without any damage resulting.'*

The differential valve-gear is further illustrated by the accompanying figures (92 to 95), which represent the gear as applied to the compound engine. 'The diagrams are not drawn to scale, but are intended to show clearly the action of the gear. The main slide-valve, G, is actuated by the piston-rod, through a lever, H, working on a fixed centre, which reduces the motion to the required extent and reverses its direction. The valve-spindle is not coupled direct to this lever, but to an intermediate lever, L, which is jointed to the first lever, H, at one end; the other end, M, is jointed to the piston-rod of a small subsidiary steam-cylinder, J, which has a motion independent of the engine-cylinder; its slide-valve, I, being

* *Engineering.*

DIAGRAMS

Illustrating Action of Differential Valve Gear

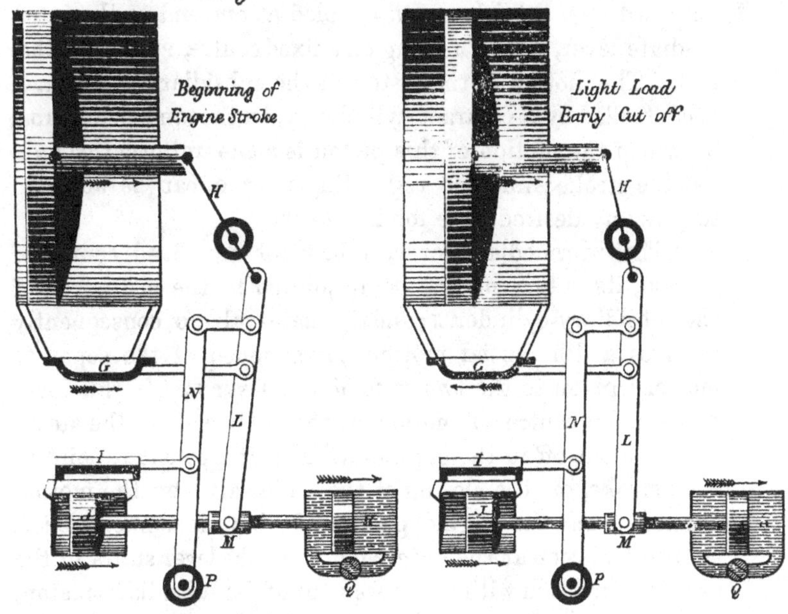

FIG. 92.

FIG. 93.

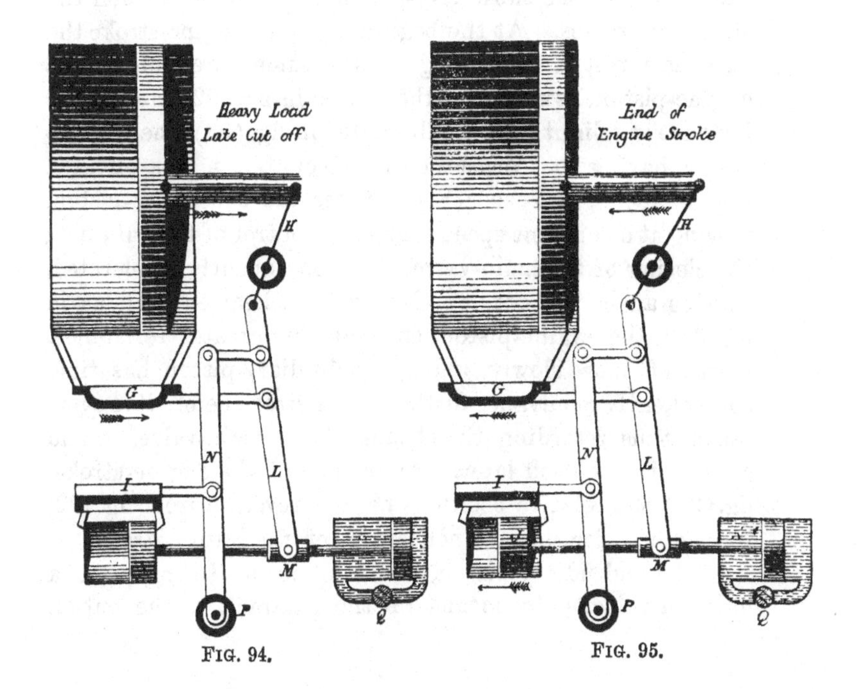

FIG. 94.

FIG. 95.

actuated by a third lever, N, coupled at one end to the intermediate lever, L, and moving on a fixed centre, P, at the other end. The motion of the piston in the subsidiary-cylinder, J, is controlled by a cataract-cylinder, K, on the same piston-rod, by which the motion of this piston is made uniform throughout the stroke; and the regulating-plug, Q, can be adjusted to give any desired time for the stroke.

'The intermediate lever, L, has not any fixed centre of motion, its outer end, M, being jointed to the piston-rod of the subsidiary-cylinder, J; and the main-valve, G, consequently receives a differential motion compounded of the separate motions given to the two ends of the lever L. If this lever had a fixed centre of motion at the outer end, M, the steam would be cut off in the engine-cylinder at a constant point in each stroke, on the closing of the slide-valve by the motion derived from the engine piston-rod; but, inasmuch as the centre of motion at the outer end, M, of the lever shifts in the opposite direction with the movement of the subsidiary-piston, J, the position of the cut-off point is shifted, and depends upon the position of the subsidiary-piston at the moment when the slide-valve closes. At the beginning of the engine-stroke the subsidiary-piston is moving in the same direction as the engine-piston, as shown by the arrows in fig. (92); and in the instance of a light load, as illustrated in fig. (93), the engine-piston having less resistance to encounter, moves off at a higher speed, and sooner overtakes the subsidiary-piston, moving at a constant speed under the control of the cataract; the closing of the main-valve, G, is consequently accelerated, causing an earlier cut-off. But with a heavier load, as in fig. (94), the engine-piston, encountering greater resistance, moves off more slowly, and the subsidiary-piston has time consequently to advance further in its stroke before it is overtaken, thus retarding the closing of the main-valve, G, and causing it to cut-off later. At the end of the engine-stroke, fig. (95), the relative positions become reversed from fig. (92) in readiness for the commencement of the return stroke.

'The subsidiary-piston, J, being made to move at a uniform velocity by means of the cataract, K, the cut-off consequently takes place at the same point in each stroke, so long as the engine continues to work at a uniform speed; but if the speed of the engine becomes changed in consequence of a variation in the load—if, for instance, the load be reduced, causing the engine to make its stroke quicker—the subsidiary-piston has not time to advance so far in its stroke before the cut-off takes place, and the cut-off is therefore effected sooner, as in fig. (93). On the contrary, if the load be increased, causing the engine-stroke to be slower, the additional time allows the subsidiary-piston to advance further before the cut-off takes place, and the cut-off is consequently later, as in fig. (95). This adjustment of the cut-off point in accordance with each variation in the load is entirely self-acting, and takes places instantly, however sudden or extensive the variation in the load on the engine may be; consequently, the engine is rendered safe in working against variable loads, as it automatically and instantly varies the distribution of steam with every increase or decrease of the resistance.

'The force acting on the subsidiary-piston, J, is much greater than that required for moving the slide-valve, the excess being absorbed in driving the fluid in the cataract-cylinder, K, through the small adjustable aperture, Q; and as the resistance of the fluid increases as the square of the velocity, a very small variation only in the speed of the subsidiary-piston can be effected by a considerable variation in the force upon it; so that the speed is maintained practically constant for a given adjustment of the cataract-plug, Q, although the boiler-pressure of steam may vary. The main slide-valve, G, is opened at the beginning of each stroke by the motion of the subsidiary-piston, which is controlled by the cataract; and a pause is consequently given at the completion of each single stroke of the engine, which allows time for the pump-valves to fall to their seats. Slip in the water is by this means prevented, as well as the shock which occurs when pump-valves close under the pressure from a moving plunger. This freedom from shocks in the pumps is an important point, giving safety from accidents, such as the bursting of pipes, and at the same time the durability of the valves and seats is materially increased.

'A plan of the differential valve-gear is represented in fig. (96): R is a connecting-rod which, with the long lever, H, fig. (92), communicates the motion of the engine-piston to the intermediate lever, L, of the valve-gear. An independent steam-starting or pausing-cylinder is shown at S, the piston-rod of which carries a rack-gearing into a pinion on one end of a tubular shaft, T; the other end of the shaft being made with a screw thread, its rotation traverses the outer end, P, of the lever, N, of the valve-gear, and thereby opens the small slide-valve, I, of the subsidiary-cylinder, J. The slide-valve of the pausing-cylinder, S, is moved by tappets by means of a lever which is actuated by the same lever, H, that works the inner end of the intermediate lever, L, of the valve-gear. The pausing-cylinder, S, is itself also controlled by a cataract-cylinder on the same piston-rod, and the length of pause after each stroke of the engine is consequently determined jointly by this and the cataract, K. The first regulates the time of opening the small slide-valve, I, for starting the subsidiary-piston, J; after which this piston, under the control of the other cataract, K, has to travel a sufficient distance for opening the main slide-valve, G, of the engine. The adjustment of the cataract, K, also determines the mean piston-speed of the engine under the normal conditions of load and steam-pressure.'*

From the foregoing description of the valve-gear it will be understood that every erratic motion of the engine alters the relative position of the valves with respect to the main-piston, and in that way the engine checks itself. So perfect is the action of this gear that when properly adjusted the full load may be thrown suddenly off the engine without any injury resulting. The effect of a sudden loss of load is to reverse the action of the valves, and *to throw the steam against the motion of the piston, stopping it before the end of the stroke.* Many instances of this have occurred in practice

* 'Direct-acting Pumping Engines and Pumps, by Henry Davey, M.I.M.E. (*Proceedings Inst. Mech. Engineers*, Oct. 29, 1874, pp. 261-264.)

FIG. 96

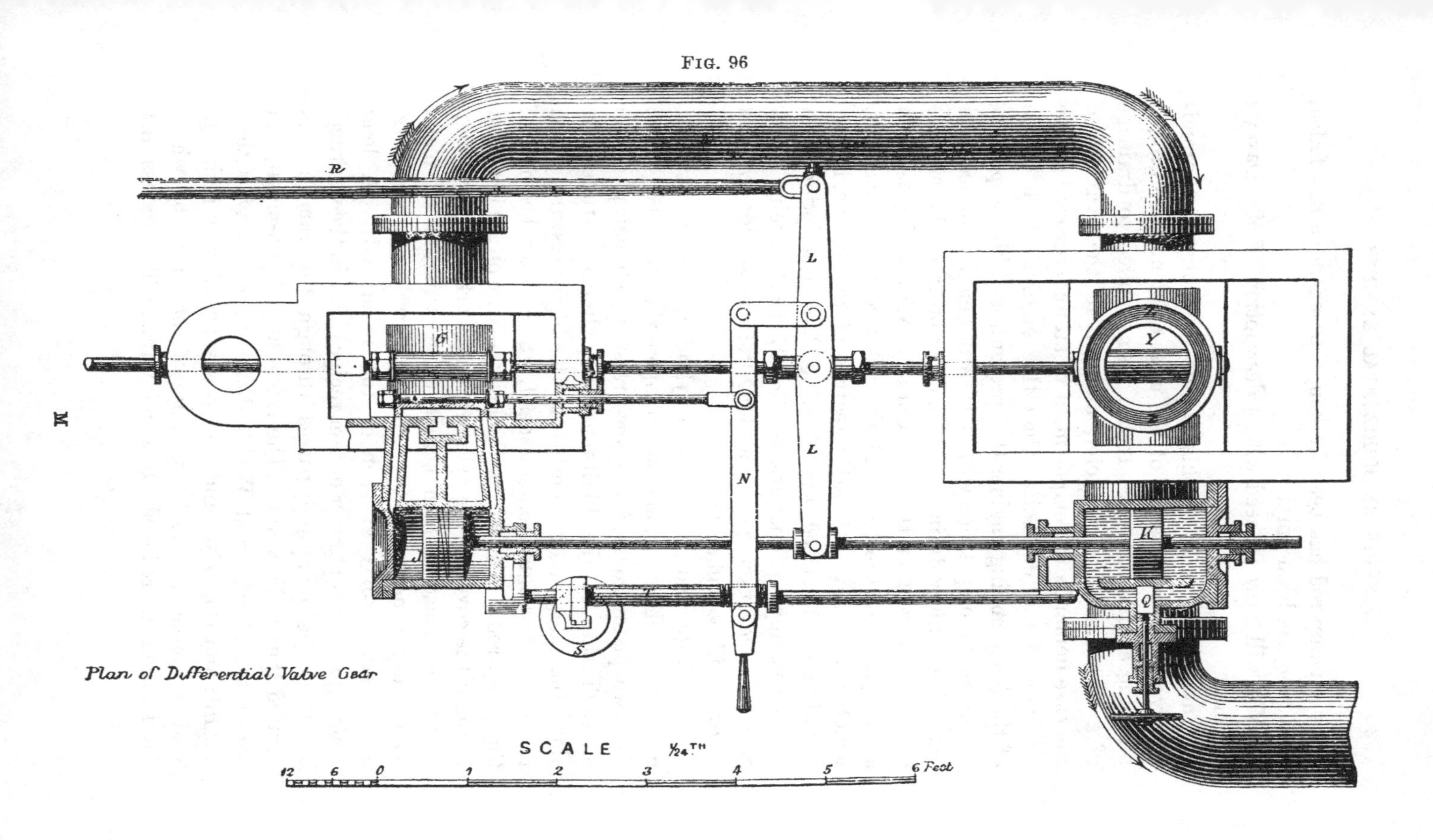

Plan of Differential Valve Gear

when a pump-rod has been broken, a pump-valve has failed, or a pipe has been burst.

We will now proceed with a description of Mr. Davey's engine.

Introduced in 1874, this engine has come very prominently to the front as one of the most efficient underground pumping engines in use. It has been applied to drainage of a very extensive character, and where the exigency of the case demands the employment of an underground engine we know of nothing better than this engine, which is one of the few pumping engines that presents anything approaching a serious competition with the Cornish engine for deep and extensive colliery drainage. As a description of the engines of this type used in the Erin Colliery, Westphalia, is given a few pages further on, and the 'Differential' valve-gear, which is the most noticeable feature of the engine, having already been described, the following short description, with the accompanying illustrations, figs. 97, 98, 99, being a longitudinal elevation, plan and longitudinal section of this engine, will be sufficient:

The high and low-pressure cylinders are in a straight line with each other, and stand upon the same bed-plate. The two piston-rods of the low-pressure cylinder pass one on each side of the high-pressure cylinder, in tubes cast along the sides, and are united to the high-pressure piston in a strong wrought-iron crosshead, which also takes hold of the pump-rod. A rod from the low-pressure piston is carried through the cover of the cylinder, and works the air-pump of the condenser, which stands on a separate bed-plate. The pump consists of two working-barrels, or ram-cases, with a stuffing-box at each of the inner ends. A long hollow gun-metal plunger or ram reciprocates in the working-barrels. The pump-rod passes through the ram, and is secured thereto by a nut. The valves are of gun-metal, and are of the Harvey and West's double-beat variety, with a low lift, and the valves and valve-seats are of such a size as to allow ample area for the passage of the water, by which the friction is reduced to a minimum. The valves lift

Fig. 97.

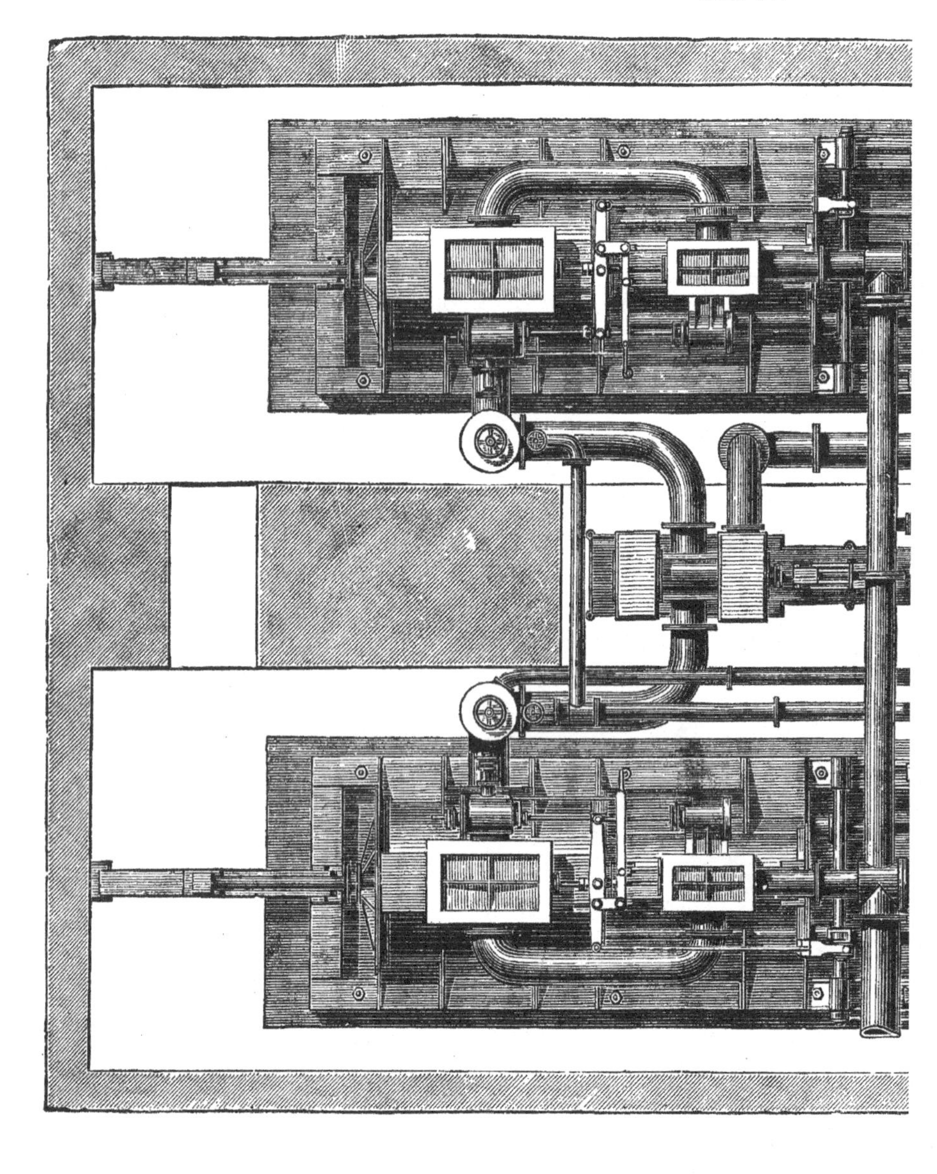

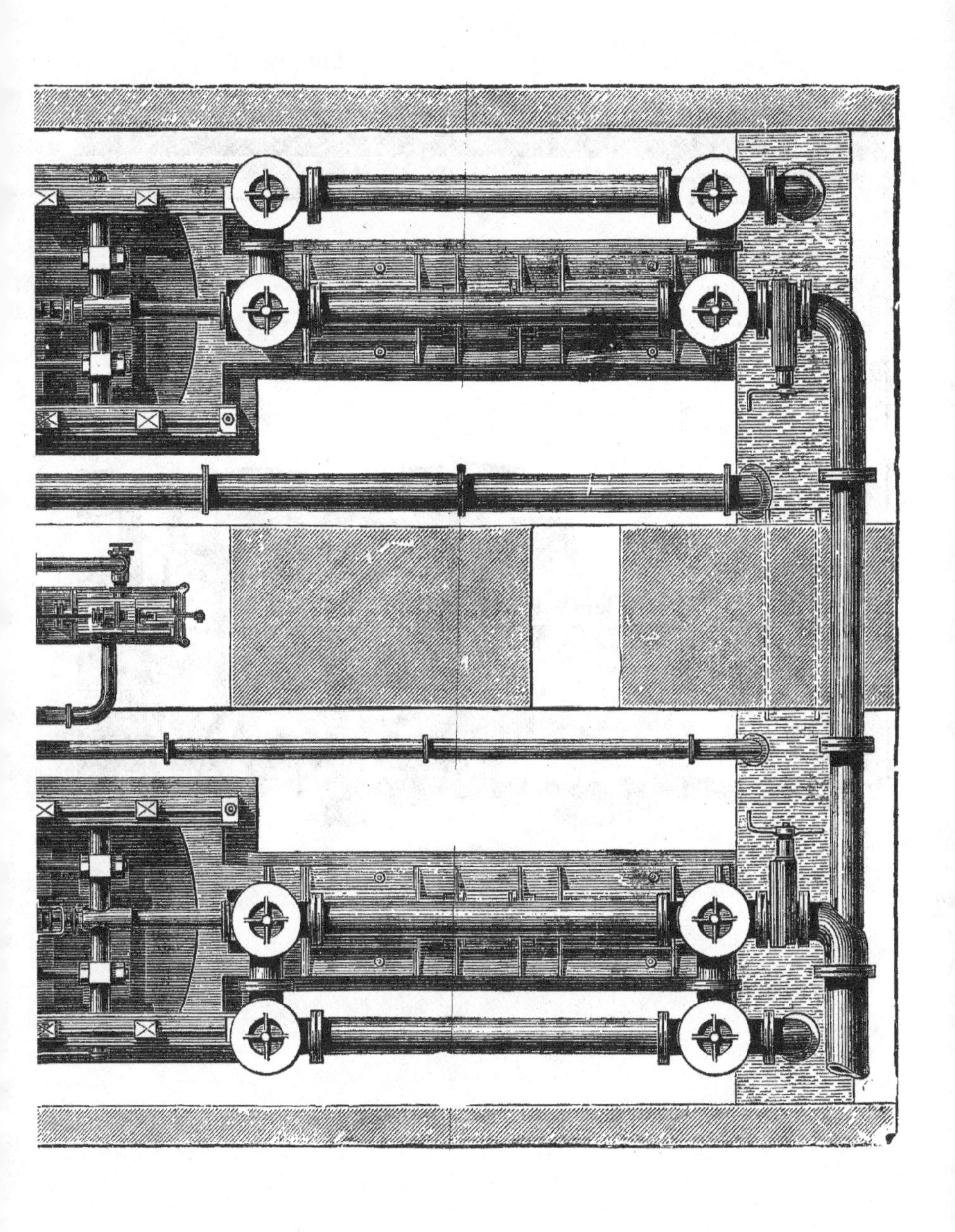

Fig. 98.

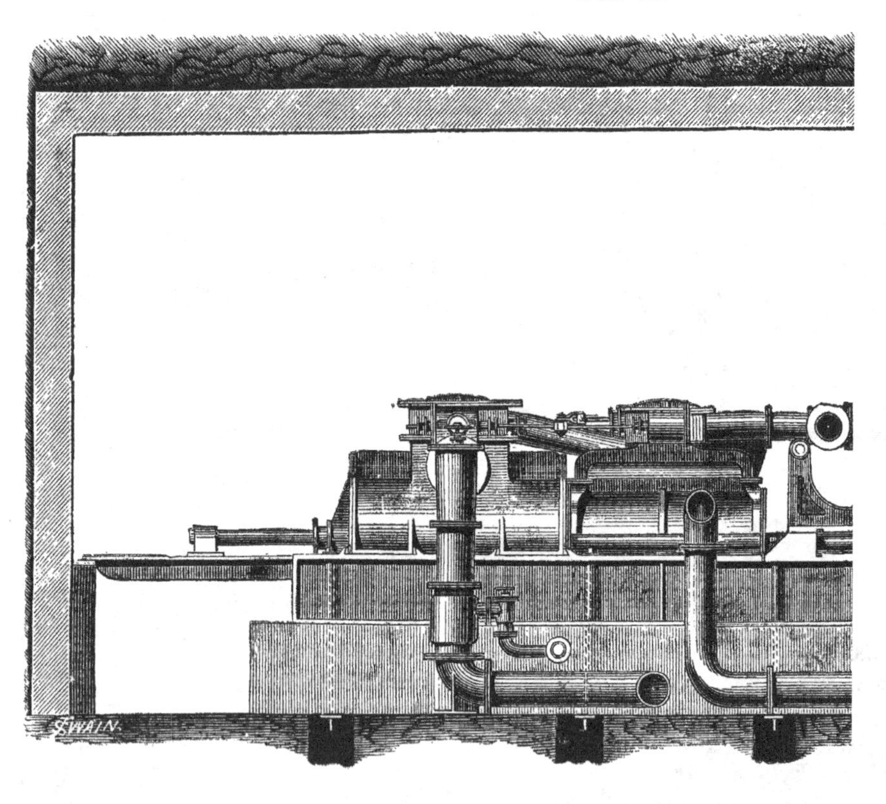

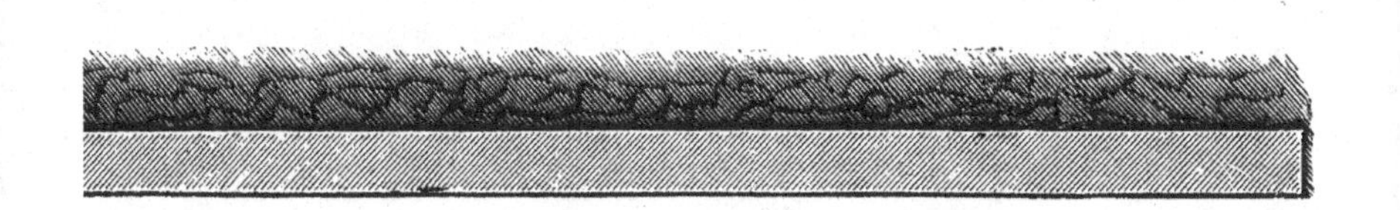

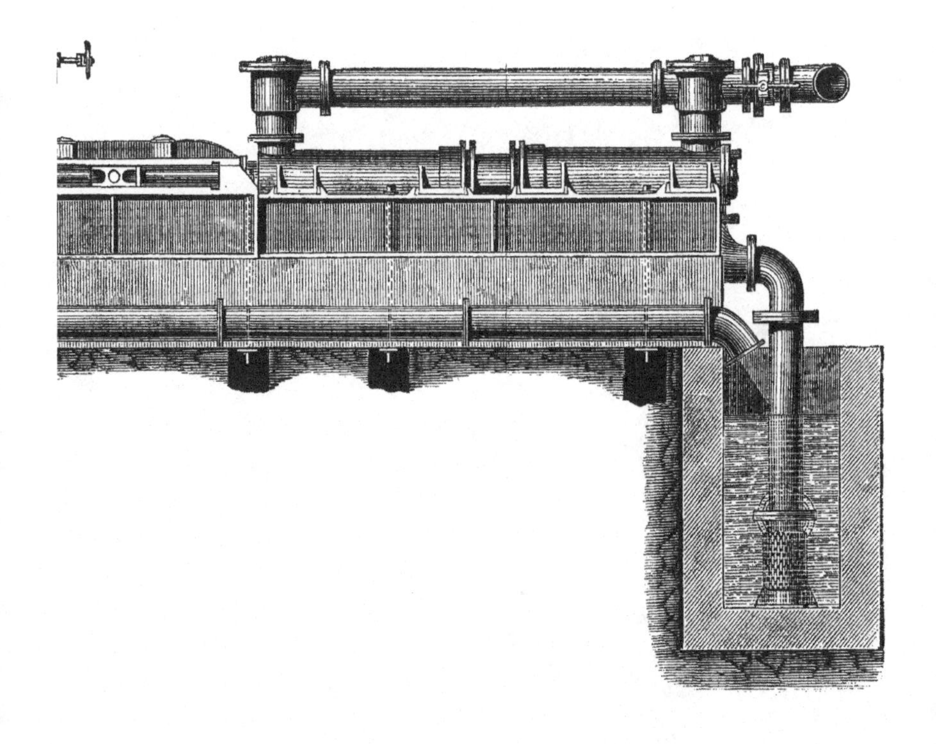

against an india-rubber spring or buffer, which assists their closing rapidly and without any shock.

The suction valve-chests are attached one to each of the outer ends of the working-barrels, and the delivery valve-chests are fixed directly above them. Davey's patent separate condenser is used in connection with this engine; by its application the makers claim that from 30 to 40 per cent. increased power is available, and a similar percentage saved in fuel. The air-pump, which is double-acting, is worked by a small 'differential' steam-engine. The air-pump valves are rubber disc-valves on gun-metal seats. By the use of this condenser the vacuum is maintained unimpaired when the engine is irregular in its working.

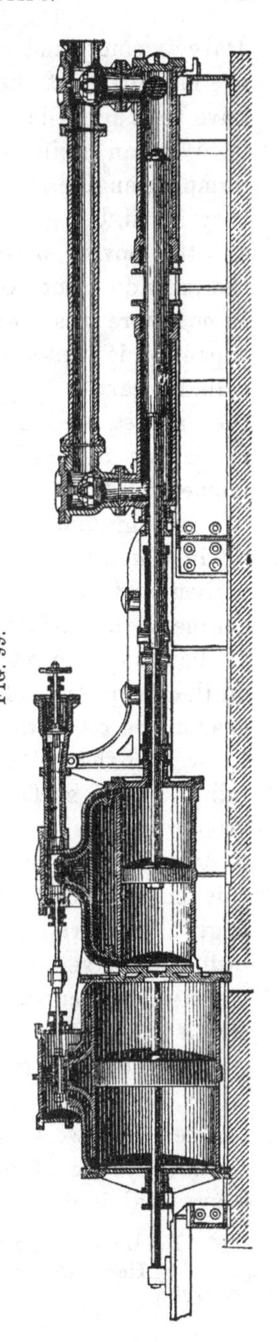

FIG. 99.

An adjunct to this engine is Davey's Patent Differential Hydraulic Pumping Engine, which precedes the main-engine, and entirely removes the danger of its being flooded. The hydraulic-engine is placed at the bottom of the pit, and it will work under water, and can be actuated from the main-engine room. The mode of its application has already been explained, and in the article on 'Hydraulic Pumping Engines' a description of the engine will be given.

The particulars given at pages 134, and 165 to 169, of work done by

Davey's compound engines will be interesting as showing the magnitude of the drainage to which engines of this class have been applied:

With an engine of this type, and indeed all underground pumping engines, the mass of matter to be set in motion is very small, being only the weight of the column of water and the moving parts of the engine. The total weight of the mass does not exceed $1\frac{1}{2}$ times the water-load; it usually is not more than about $\frac{1}{10}$ or $\frac{1}{8}$ greater. The ratio of expansion is consequently not high, and the high-pressure cylinder carries very nearly full steam. The expansion in the engines used in the Erin colliery is about 3 to 1, which is all that is possible to get with the available inertia and momentum.*

Underground engines cannot be used for pumping out a pit in which the operation of sinking is being carried out without additional pumping apparatus. Mr. Davey places his engine at the surface and connects it to a pair of quadrants at the top of the pit, which are attached to pump-rods, and on the completion of the sinking operations the pumping-gear in the pit is thrown aside, and the engine is removed to or near the bottom of the pit. The engine can be retained at the surface to continue the drainage of the pit after it is sunk, should there be any objection to placing it underground. For pits already sunk to the required depth, this engine possesses some advantages over the Cornish engine. It permits the winding to be done in the same pit, owing to the very small space which the water column and steam-pipes occupy. The water is raised in one direct lift, which is a much simpler arrangement than the Cornish lifts, it does not require so much attention, and is far less costly, and the engine is very much cheaper than the Cornish engine. The working economy, however, is in favour of the Cornish system of drainage, and in metalliferous districts where coal is dear this is a very important consideration.

* 'The Underground Pumping Machinery at the Erin Colliery, Westphalia,' by Henry Davey, M.I.C.E., &c. (*Trans. Society Engineers*, June 12, 1876.)

Under such circumstances the underground engine can scarcely be said to offer a serious competition either to the Cornish engine or to Mr. Davey's surface compound engine. Many of the forms of pumping machinery we have already discussed have some particular sphere of operation in the mining field for which they have a special adaptation. This engine is peculiarly adapted for use in colliery pits already sunk, in which it is necessary to replace worn-out machinery and to enable winding to be done in the same pit, or to supersede the pump-lifts used during the course of sinking, when the whole of the pit is wanted for winding.

This engine appears to give satisfactory evidence of durability. The engines used at the Erin Colliery (up to June 1876) had been working for nearly two years, night and day, without stopping, except for the grinding-in of the pump-valves when they got indented by grit or from any other cause. That occurred, perhaps, once in two or three months. The pistons had not perceptibly worn.* Impure water, or water containing grit, affects the rams and valves more than any other parts. Cast-iron valves become useless in a few days, and brass valves after a few weeks generally require refacing. These difficulties have led to the adoption in some instances of valves compounded of hippopotamus hide and brass, which give much better results.

The following is an account, extracted from the 'Trans. Soc. Engineers,' June 1876, of Davey's Differential Compound Engines, used at the Erin Colliery, Westphalia; it is perhaps the largest example of underground pumping machinery extant:

'The pit is 1,200 feet deep, and the pair of steam-engines and hydraulic-engines combined are together capable of delivering the required quantity of 1,000 gallons of water per minute, or 60,000 gallons per hour, to the surface. At a point 900 feet below the surface are placed a pair of compound differential pumping engines, as shown in the section (fig. 89) with separate condensers. Each engine is capable

* *Trans. Soc. Engineers,* June 1876 : Davey's Account of the Underground Pumping Machinery at the Erin Colliery.

of raising 500 gallons per minute to the surface, and at the same time of supplying power to a pair of hydraulic pumping engines placed at the bottom of the pit, and employed in raising 500 gallons each per minute to the main-engine. Each engine has a massive girder bed of cast-iron, with a 35-inch high-pressure and a 60-inch low-pressure cylinder, 6-feet stroke. The normal speed of the engine is 12 strokes per minute. The slide-valves for these cylinders are actuated by the author's patent differential valve-gear. The pistons are fitted with metallic packing. The pumps are double-acting ram-pumps, 12½ inches diameter, 6 feet stroke, with rams of gun-metal. The pump-rods, 5¾ inches diameter, pass from end to end of the rams, and are secured thereto by gun-metal nuts. The piston-rods and pump-rods are united in one strong wrought-iron crosshead, supported and guided in cast-iron slide-bars. The pump-valves are of gun-metal, double-beat, 11 inches in diameter, and low lift, allowing ample area in the water-passages, and thus reducing friction to a minimum. The suction-pipes are 10 inches diameter, and the rising-main 14 inches diameter; all the joints of the pumps and pipes are of the well-known hydraulic type, made with gutta-percha.

'Each pump is provided with a sluice-valve on the delivery side, 10 inches in diameter. There is one condenser to each engine. The steam-cylinder is 14 inches diameter and 3 feet stroke, working a double-acting air-pump 16 inches diameter and 3 feet stroke. The air-pump valves are rubber disc-valves on gun-metal seats. Each condenser is of sufficient capacity for both engines, and the connections are so arranged that either or both engines can work from either or the same condenser, thus allowing one condenser to be under repair if necessary. By the application of the separate condenser the vacuum is maintained unimpaired when the engine is irregular in its working.

'At the bottom of the pit, 300 feet below the main-engines, are placed a pair of hydraulic engines (see remarks on "Hydraulic Pumping Engines"). These engines lift 1,000 gallons per minute to the main-engine,* at a speed of 12 strokes per minute, whilst the main-engines finally force the water to the surface, and at the same time, through the column, supply power to work the hydraulic-engines.† The hydraulic-engines consist of double-acting power rams, and pumps placed face to face on the telescopic principle. These engines are self-contained on a strong girder bed similar to that of the steam-engine. The pumps which are placed at each end of the bed are 14 inches in diameter, and of the ordinary ram type, with gun-metal double-beat valves, and brass-covered rams 12 inches diameter and 5 feet stroke. The two rams are connected to each other by two strong wrought-iron side-rods, and thus alternately pull one another in their return stroke. The power rams, which are 7 inches diameter, are wholly of gun-metal, and of the same stroke as the pumps. They are connected at one end to a central valve-box, and being cast hollow admit the pressure to the cylindrical rams of the pumps. All the bearings and glands are brass-bushed. All the packings are easy of access, and are applied only in stuffing-boxes.

'The main-valve box is placed in the centre of the

* 'But it is necessary for the main-engines to lift more than that quantity, because there are 1,000 gallons of water as a feeder at the bottom of the mine, and it is required to deliver that quantity on the surface. If 1,000 gallons only be lifted by the main-engines, the quantity delivered to the surface would be only the difference between the quantity lifted by the hydraulic-engines, and that required to work them. There is a certain quantity of water made available in the rising-main for working the hydraulic-engines, and in order to provide that quantity, it is necessary to make the pumps on the main-engines so much larger. The main-engines really pump to the surface 1,400 gallons per minute, or what is equivalent to it. They pump it against the head due to the distance from the engines to the surface. 1,000 gallons are pumped to the surface, and the other 400 gallons pass down and act simply as a medium—a water spear in fact.'

† 'The available head for driving the hydraulic-engines is the head of water above the main-engines; this is the effective head. The head between the main-engines and the hydraulic-engines being the same, the two columns of course balance each other, and as far as the supply is concerned the water that goes down is simply that which gives power to the hydraulic-engines.'

engine, and contains four valves, 5¾ inches diameter, two being supply, and two exhaust, which communicate with their respective pumps. The valves are cylindrical, gun-metal, mitre valves, 6½ inches diameter at the top, and provided at their cylindrical ends with double cup-leathers. They are of large diameter and low lift, thus reducing friction and wear and tear to a minimum. The method of actuating these valves has been patented by the author, and consists in lifting and closing the valves by water-pressure. Thus, in the arrangement illustrated (see remarks on "Hydraulic Engines"), the upper or cylindrical portion of the valve is of rather larger diameter than the valve proper, consequently the pressure from the main lifts this valve, and admits pressure to the power-ram. On the return stroke this valve is closed by admitting pressure on the large area at the top of the valve, whilst a corresponding but intermittent action takes place with the relief-valve. The main-valves thus rise and fall by the alternate admission and cutting off the water-pressure, which is further regulated by a modification of the differential gear applied to the two small change valves. One end of the intermediate lever is actuated by the hydraulic-cylinder, which is 2½ inches diameter and 8 inches stroke, and has an independent motion, regulated by a small slide-valve in the change-box, whilst the other end of the lever is worked by suitable rods and levers from the engine itself. The small valve contained in the change-box admits or relieves the pressure upon the main-valves. The slide-valve for the cataract-cylinder has no lap, but the slide-valve for the main-valve box has a negative lap; the length of time therefore governed by the cataract occupied by the small slide-valve in traversing the negative lap, gives a pause to the engine at each end of the stroke, as both supply and exhaust-valves are closed until the small slide-valve travels a certain space. This pause allows time for the pump-valves to fall to their seats. Slip in the water is by this means prevented, as well as the shock which occurs when valves close before a moving plunger.

'The main and hydraulic-engines are quite independent

in their motion. That is to say, that the strokes are not identical or synchronal. The hydraulic-engines work from a reservoir of power, that reservoir being the rising-main from the main-engines, and as long as there is water in the main, the hydraulic-engines may be worked at any speed within the maximum. By this hydraulic system of underground pumping the main-engines are kept out of danger of flooding. The hydraulic-engines will work under water, and can be actuated from the main-engine room. As a further security they could be placed in a water-tight staple, and by such means the hydraulic-engines could be under repair even when the water rose to the main-engines 300 feet up the shaft.'

At the Nixon's Navigation Colliery, South Wales, the author inspected one of Mr. Davey's engines at work underground 365 yards (about 1,100 feet) from the surface. It had a high-pressure cylinder of 28 inches, and a low-pressure cylinder of 50 inches, with a 4½-feet stroke, working a double-acting ram of 12 inches diameter. The suction and delivery-pipes were 10 inches diameter. The steam is taken down the pit, and along a heading, to a receiver at the back of the chamber in which is the Davey engine, from the surface in a 12-inch pipe; a branch pipe of 6 inches supplies the engine, the remainder of the steam being used for working two pairs of hauling-engines (26 inches and 24 inches) and a blocking-engine (12 inches). It was used at a pressure of from 40 to 43 lbs., and the engine kept the water at 4½ strokes per minute. The water is taken from a lodge-room about 220 yards from the surface. At the time of his visit there was a hydraulic-engine of 4½-inch cylinder, ready fixed on the same floor as the large engine, and designed to take the water from a lower point, but it had not yet been used. The suction and delivery were 4 inches; the ram was double-acting, in a chamber 6 inches in diameter. The water would be discharged into the suction-pipe of the large engine. The actual horse-power of this engine is 160; equal to raising 500 gallons per minute 1,100 feet high at 10 strokes per minute. Weight of the engine and pumps 46 tons. Messrs. Hathorn, Davey, and Co., Leeds, are the makers.

THE 'SELF-GOVERNING' COMPOUND ENGINE.

The valve-gear of this pump is precisely the same as that of the simple self-governing engine, made by the same firm. There is a high and low-pressure cylinder. The two steam-cylinders and pump-cylinder are placed in a line, the back end of the high-pressure cylinder forming the front end of the low-pressure cylinder. There are two piston-rods to the low-pressure piston, which work through stuffing-boxes, one on each side of the high-pressure cylinder; these and the high-pressure piston-rod are attached to a crosshead, actuating the pump-ram. By referring to the account of the single-cylinder self-governing engine (page 115), illustrations and a description of the valve-gear will be found, which is the same in both engines. Messrs. Hayward Tyler and Co. are the makers.

CHERRY'S PATENT COMPOUND ENGINE.

This pump consists of one high-pressure steam-cylinder, contained within a low-pressure steam-cylinder. The latter has two piston-rods in a horizontal line with the high-pressure piston-rod, and the three are attached to the same crosshead, which also carries the pump-rod. The valve-chest is fixed upon the low-pressure cylinder. The slide-valve is of the compound D type, and is actuated by a pair of shooting-pistons fitted in a piston-chamber above the slide-valve. Midway between the shooting-pistons is a small slide-valve working transversely over a flat face, by means of which the steam is admitted to the back of each shooting-piston alternately. The face for the starting-valve forms part of the distance-piece of the shooting-pistons. A stud in the bottom of the distance-piece is let into the back of the main-valve, having an exhaust-passage through it, thereby establishing a communication between the shooting-pistons and the main-exhaust. On each side (in transverse line) of the exhaust-passage referred to, is a longitudinal steam-passage, which communicates, through the distance-piece, to each end of the piston-chamber.

The starting-valve is actuated by a lever, anchored either to the crosshead or to the distance-bar, between the steam and pump-cylinders.

Two or three types of water-cylinders or pumps are applied to this engine, according to the height of the lift, and the purpose for which it is required.

When this engine is required condensing it is fitted with an air-pump condenser, working behind the steam-cylinders, or, when advisable, with Holman's Patent Condenser.

*** Since the above was written Messrs. Tangye Brothers and Holman, the makers of the engine, have informed the author (December 1879) that they do not now make this type, the double-cylinder compound engine, noticed at page 181, having taken its place.

Sturgeon's Patent Compound Engine.

This engine consists of one low-pressure steam-cylinder in the centre line of the pump, combined with a pair of high-pressure cylinders, one on each side of the low-pressure cylinder. The three piston-rods are attached to a common crosshead; the middle rod is continued through to the pump and takes hold of the pump-rod. The condenser is attached to the suction-pipe of the pump.

The valve-box is placed upon the top of the low-pressure cylinder. The steam-valve is of the compound D type, the larger enveloping the smaller, and the semicircular space between the two backs forming a steam-passage, through which the used steam from the high-pressure cylinders is conducted to the low-pressure cylinder. The valve works on a face with a double set of steam-passages, and an exhaust-branch in the centre, the outer ports communicating, through ports in the cylinder-end covers, with the two high-pressure cylinders, and the inner ports passing direct into the low-pressure cylinder. Thus the used steam from the high-pressure cylinders is conducted back through the valve from the ends of the high-pressure cylinders to the opposite end of the low-pressure cylinder, finally passing through the exhaust and into the condenser. A pair of small starting-valves work one on each end of the valve-face; the spill connecting them passes out through a stuffing-box, and is attached to a tappet-lever, which is actuated at each end of the stroke by a projection on the crosshead. The office of these valves is to furnish steam and exhaust communication to a valve-moving piston in the upper part of the valve-box, which in turn gives the requisite motion to the main-valve.

The engine is connected to two water-cylinders fitted with a long ram. The piston-rods of the two high-pressure cylinders are attached to a collar on the pump-ram, or rams, as they are actually, for the long ram is divided by the collar into two rams, each working in a cylinder, thus imparting to them the motion of the engine. An engine of this construction, with 26-inch low-pressure and 11-inch high-pressure cylinders, 7-inch rams, with a 2-feet stroke, capable of raising 10,000 gallons per hour 600 feet, measures, over all, 13 feet 8 inches long, 5 feet 2 inches wide, and 8 feet to the top of the air-vessel. Engines of this type have been made for Messrs. Thomas Fletcher and Sons, Outwood and Stopes Colliery, near Manchester.

The following are the special advantages which this engine is stated to possess: 1st. 'A great saving in length, as compared with end-to-end cylinders. This is a most important point in mining pumps for underground working, while for other purposes it also enables a considerable saving in foundations. 2nd. Greater simplicity of parts; a single-valve and valve-gear serving the purpose of two distinct and separate valves and valve-gears on the old system, and, 3rd, is consequently more easy to keep in repair. 4th. Saving in power by avoiding the loss of friction, leakage, &c., attendant upon the extra valve and valve-gear required on the old system. 5th. Considerable saving in power, owing to the immediate and short passage of the steam from the high to the low-pressure cylinders, the loss in pressure, owing to the steam having to fill the long pipe and extra valve-box in the end-to-end cylinder arrangement being avoided. 6th.

Saving in steam in its passage to the low-pressure cylinder, by reason of the port through which it passes being "jacketed" by the high-pressure steam outside it the greater part of the way, and heated the remainder by reason of the high-pressure steam having passed through the port in the previous stroke. On the old system, the ports are not only not "jacketed," but are cooled by the passage of the exhaust-steam through them at every stroke.'

The makers of this engine are Messrs. John H. Wilson and Co. Sandhills, Liverpool; also makers of the 'Selden' Steam-pump.

*** Since the foregoing was written, the author has been informed by the makers that they have given up the manufacture of this engine, but the arrangement being an unusual one, and considering that failures and successes are not without their useful lessons, he has decided to let his remarks remain, with a slight abridgment.

Parker and Weston's Patent Compound Engine.

The accompanying figs. 100 and 101, are views of Messrs. Parker and Weston's compound engine, with an 8-inch high-pressure, and a 12-inch low-pressure cylinder, and an 8-inch piston-pump, and calculated to force 13,000 gallons per hour about 130 feet high. The high-pressure cylinder is placed behind the low-pressure, and they have a piston-rod common to both, which is also connected to the pump-piston in the same line. Each cylinder has a separate steam-chest and set of valves receiving independent motion from their respective cylinders, but connected by a small rod to prevent any possibility of their not moving together.

The following particulars are supplied by the makers, Messrs. The Coalbrookdale Company:

'The cylinder-covers and distance-pieces are so arranged that by merely unscrewing the nut at the back of the pump-piston, and by moving the left-hand cover of each steam-cylinder, the pistons may be slipped out, and the rings examined or replaced without disturbing the cylinders

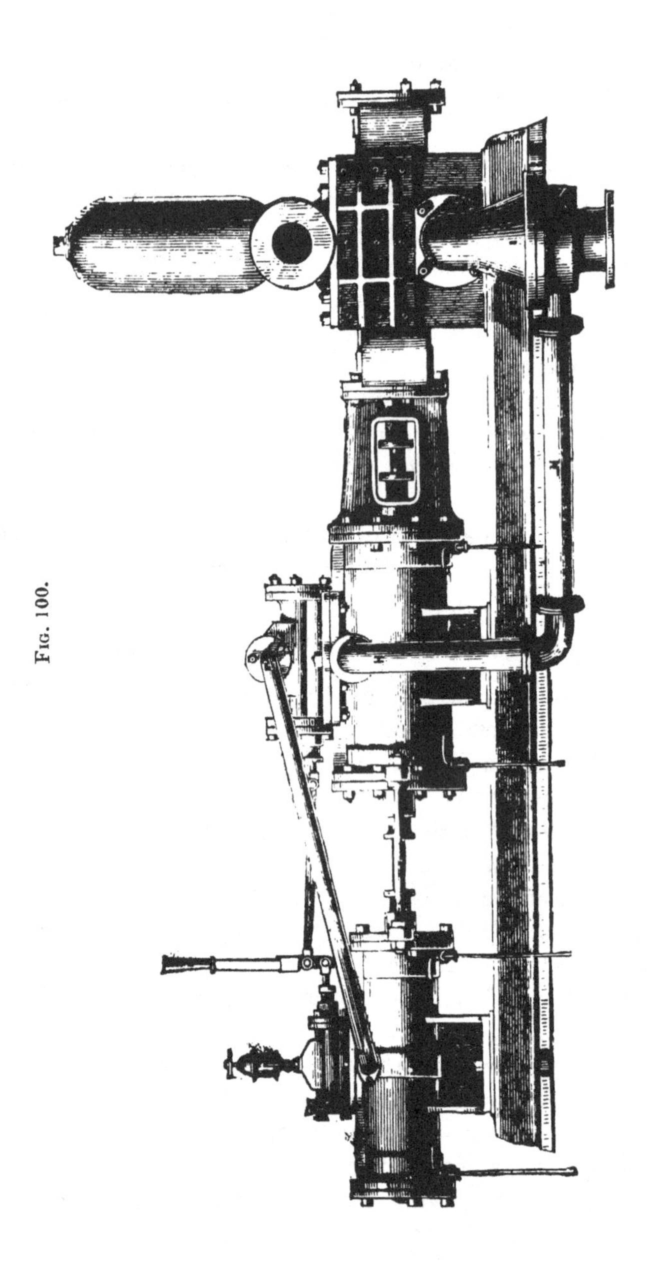

Fig. 100.

themselves. Where sufficient space can be obtained for fixing the engine, these compound pumps are found to answer admirably for the purpose of saving fuel. They are made in various combinations of cylinders and pumps, and when fitted with the cataract expansion-gear in connection with a ram, prove themselves highly economical and efficient.

'The condenser shown in connection with the compound-cylinders will be found very simple, and will effectually get rid of the exhaust-steam, giving a vacuum of about 8 or 10 lbs., but should a better vacuum be sought a separate air-pump and condenser will be necessary. We give above a sectional view of the combined direct-acting air-pump and condenser as applied to high-pressure engines, thereby converting them into economical and more efficient machines. In connection with a direct-acting pumping engine, the air-pump would be worked by the piston-rod carried through the back cylinder-cover, dispensing with the extra steam-cylinder; but for the purpose for which the above form is specially designed, it is more advantageous to have it as a separate engine, since it does not necessitate the slightest alteration or stoppage of the engine to which condensation is to be applied, and it may be placed in any situation within reasonable distance of the engine, and where sufficient water can be obtained. Another great advantage in the separate condenser is that the steam from several engines may be turned into it. The steam used in the cylinder of the direct-acting air-pump is comparatively small, as a proportion of 6 to 10 is maintained between the

Fig. 101.

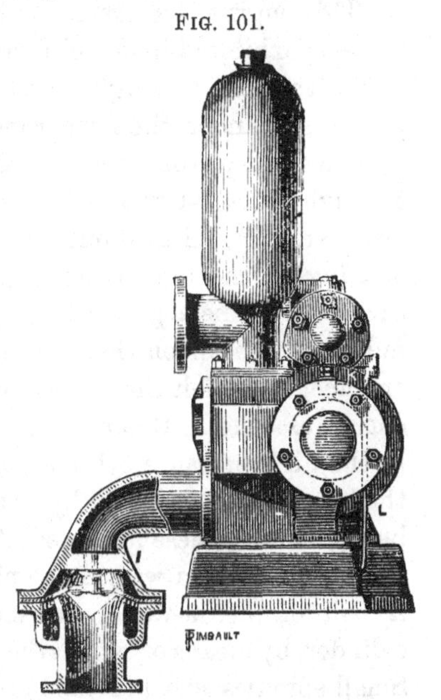

areas of engine and pump-cylinders, and the exhaust steam from the steam-cylinder being turned into the condenser, a great portion of the work is done by the vacuum.'

For a description of the valves and other details, see the account of the simple engine, page 122.

Field and Cotton's 'Direct Expansion' Compound Engine.

This engine is a novelty in the way of compound engines, and was exhibited in model form at the Vienna Exhibition in 1873. It consists of a high-pressure cylinder placed within a low-pressure cylinder, the former moving backwards and forwards upon a fixed piston attached to a piston-rod, which, passing through a steam-tight bushing in the end of the high-pressure cylinder, is fitted into one end of the low-pressure cylinder. The high-pressure cylinder, being fitted with piston-heads at each end, forms a piston for a low-pressure cylinder, the annular space between the two piston-heads forming a steam-chamber to which the live steam from the boiler has constant communication. Steam is admitted from the steam-chamber to the high-pressure cylinder, and thence is discharged into the low-pressure cylinder by a pair of piston-valves, connected by a spindle. These are fixed at the bottom of the high-pressure cylinder, and within the piston-heads before mentioned. A pair of piston-valves are mounted upon the low-pressure cylinder, by means of which the steam is exhausted therefrom. Small spindles are provided at the end of the exhaust-valve chest and at each end of the low-pressure cylinder, in a line with the centre of the steam-valve, in order to move the latter and the exhaust-valve. These will only be required in case of any irregularity in the action of these valves.

The high-pressure cylinder making its stroke within the low-pressure requires, of course, that the latter be as long as the high-pressure cylinder and its stroke. The power of both cylinders is transmitted to the pump by a single piston-rod attached to the end of the high-pressure cylinder.

The action of this engine may be briefly described as

follows: Supposing the high-pressure cylinder to have arrived at the termination of its stroke (the piston steam-valve having previously made its stroke in the same direction), live steam is admitted from the steam-chest through a small groove in the end of the low-pressure cylinder to the back of the piston steam-valve, causing it to make a stroke in the opposite direction, thereby admitting steam from the steam-chest into the high-pressure cylinder, simultaneously exhausting the steam from the opposite end into the low-pressure cylinder. The steam emitted from the steam-chest through the small groove already referred to also communicates with the back of the exhaust piston-valve, causing it, simultaneously with the steam-valve, to perform its stroke, thereby reversing the communication of the low-pressure cylinder with the exhaust to the opposite end.

Messrs. Merryweather and Sons, Greenwich Road, London, S.E., are the sole makers of this engine, of which Messrs. E. Field and F. M. Cotton of London, are the patentees.

Walker's Compound Engine.

This engine, which is of a unique description, is the design of Mr. W. Walker, and it appears to have been introduced about the year 1874. It consists of one steam-cylinder, with an annular partition fixed at the centre of its length. A barrel of considerable length, with piston-heads at each end, is fitted into the cylinder; the barrel part works steam-tight through the annular partition, the piston-heads also working steam-tight in the cylinder. The latter is formed with the valve-chamber above it. The slide-valve, which is of cylindrical form, consists of two parts, connected together by a link, with a piston-head at each end fitting the valve-chamber. Each part of the valve is formed with a D passage, which serves to establish communication for the flow of steam from the annular spaces between the piston-barrel and the cylinder to the ends of the latter. A groove at the bottom of the piston-barrel, working over a corresponding guide at the bottom of the annular partition, prevents the piston from rotating. It will be understood there is only one piston-rod. The engravings represent an elevation (fig. 102), a side elevation of the pump, with the steam-cylinder in section (fig. 103), a transverse section through the centre of the cylinder at A (fig. 104), and a similar section through the end at B (fig. 105). The action of this pump is as follows:—Steam is admitted to the central space, between the two halves of the valve, through an opening at the top of the chamber, as indicated by the arrow, and part of the steam finds its way by leakage into the end spaces of the valve-chamber. A passage, S, extends from the left-hand end space of the valve-chamber, along

FIG. 102.

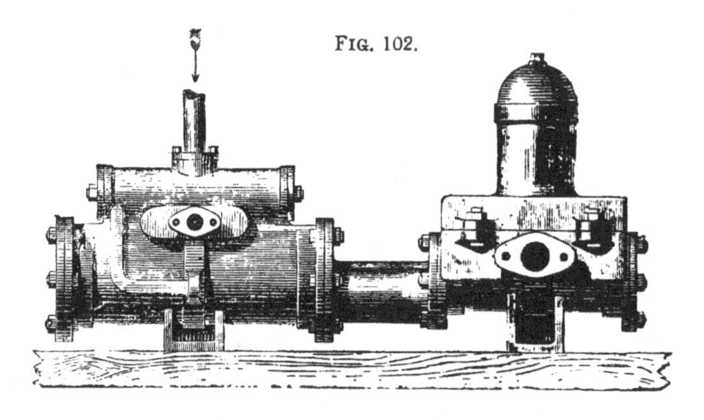

the side of the cylinder, to a port formed in the centre of the annular partition, marked *u*. A similar passage extends from the right-hand end space of the valve-chamber to a second port formed in the partition, but as the passage is formed in the part which is cut away (fig. 103) it does not appear there. Two passages are formed in the barrel, in such a manner and in such positions that when the piston approaches the termination of its stroke in either direction, one of these passages will connect one of the passages seen on the left and right of the barrel (in fig. 103) with an upper passage, *w*, which extends into the exhaust, O. (Fig. 104 shows the exhaust communication between the left-hand end of the valve-chamber, through the main-piston barrel, with the main-ex-

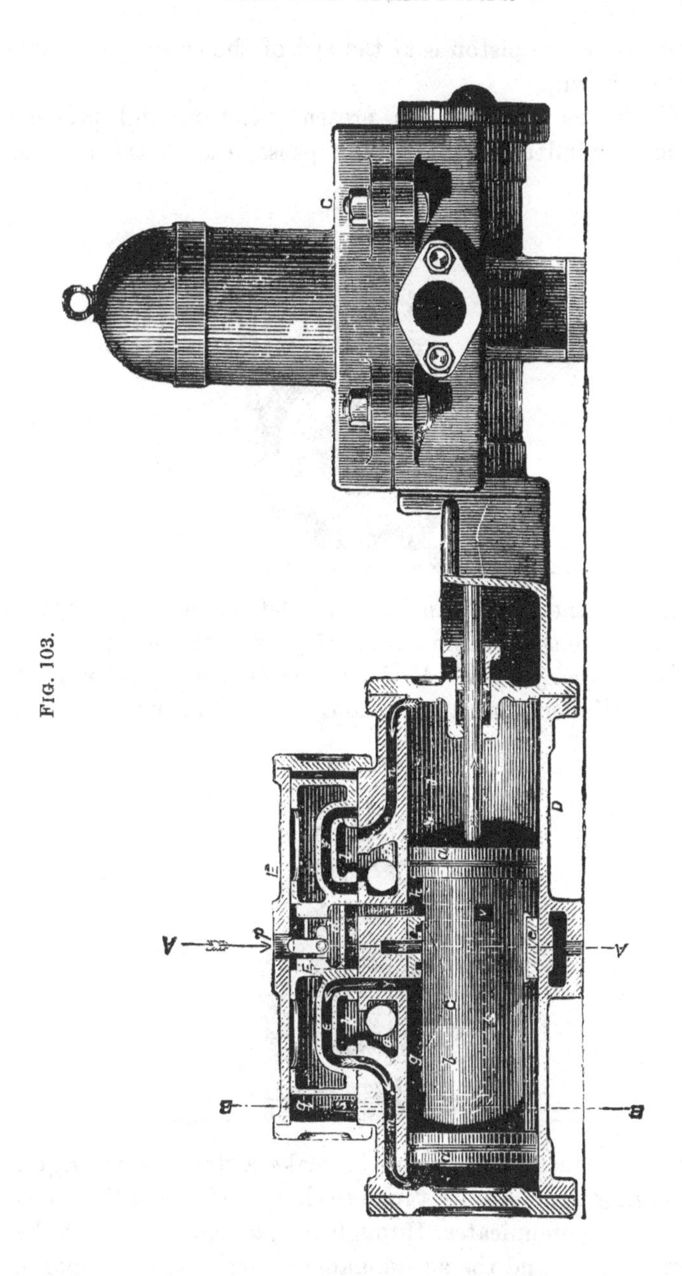

FIG. 103.

haust, when the piston is at the end of the cylinder opposite to that shown).

With the valve in its present position high-pressure steam is admitted through the passage *x* to the annular

FIG. 104.

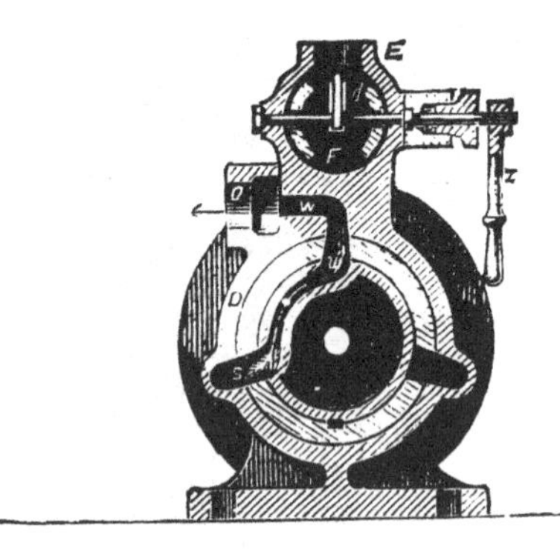

space at *h*, the used steam from the left-hand annular space, *g*, simultaneously escaping, through *y*,*e*,*m* to the left-hand end, *i*, of the main-cylinder. At the same time the right-hand end of the cylinder, *j*, exhausts through *n*, *l* to the main-exhaust,

FIG. 105.

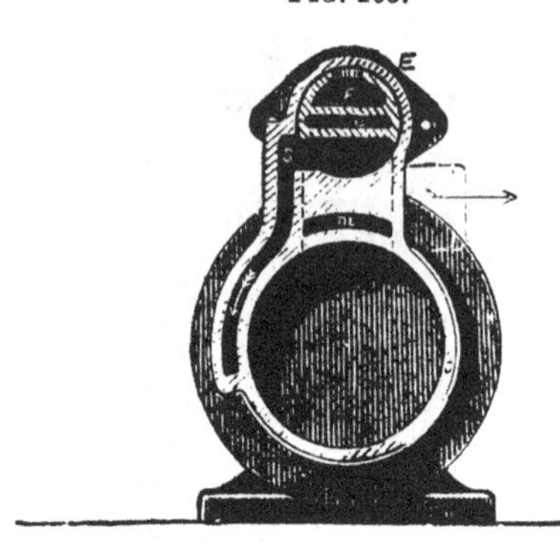

o, thereby causing the piston to make a stroke to the right. On nearing the end of its stroke the left-hand end of the valve-chamber communicates, through *s*, *u*, *w* (fig. 104), with the main-exhaust, and the accumulated pressure at the opposite end of the chamber, through leakage of steam, causes the valve to reverse to the left, thereby changing the steam and exhaust communications for the return stroke of the piston. A starting-handle is fixed on a rocking-shaft working in the centre of the valve-chamber, on which a lever is fixed, which moves the valve when the shaft is rocked by the handle.

'These pumps have been in use for many months past, and are giving every satisfaction . . . and, it is stated, effect a very marked economy of fuel.' ('Engineering,' July 16, 1875, to which the author is indebted for a description of this engine).

The makers are Messrs. Clayton, Howlett, and Venables, Atlas Works, Woodfield Road, Harrow Road, London, W. For sizes and prices see page 132.

Tangye's Compound Engine.

This engine is one of the most recent additions to our large pumping engines, having been introduced by Messrs. Tangye Brothers and Holman about three years ago. The accompanying illustrations (figs. 106 and 107) are elevations of two types of this engine, the first with a ram-pump and an air-pump condenser, and the second with a piston-pump fitted with Holman's patent condenser. The following is a description of an engine of the former type, having 21-inch and 36-inch steam-cylinders, 12-inch plungers, and a 4-feet stroke.

'The first item deserving attention is the compound special direct-acting steam-pumping engine, with condenser and with high and low-pressure cylinders, so that the steam can be used expansively to the highest degree. This arrangement has been designed for pumping water in large quantities in waterworks, collieries, and deep lifts. Whilst it is cheap in first cost, it can be worked as economically as the best class of Cornish pumping* engines, whereas the cost of the whole plant would be little more than it takes to provide the foundation of the old pumping engines.'

* This is a mistake. For a consideration of the economy of the compound engine and its comparison with that of the Cornish engine, see pages 140 to 149.

Fig. 106

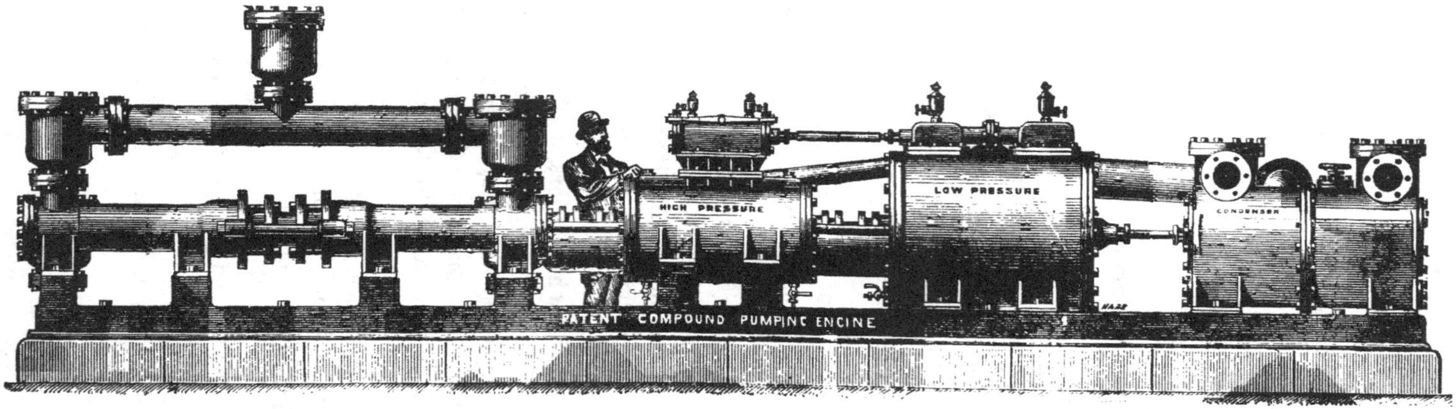

HIGH PRESSURE

LOW PRESSURE

CONDENSER

PATENT COMPOUND PUMPING ENGINE

FIG. 107.

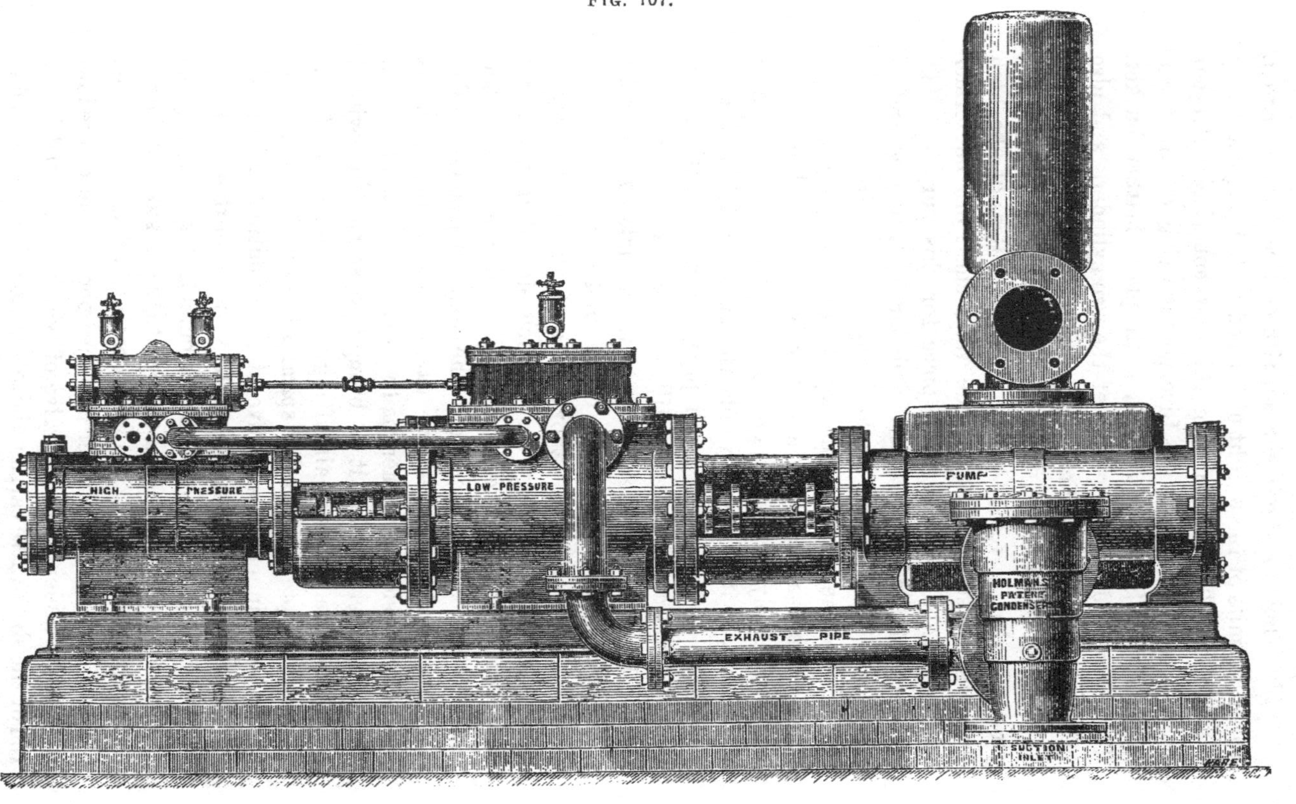
HIGH PRESSURE
LOW PRESSURE
PUMP
HOLMAN'S PATENT CONDENSER
EXHAUST PIPE
SUCTION INLET

'The whole arrangement constitutes a continuous girder, which permits of the weight of the bed-plate being materially diminished. The pump is of the new spherical valve-box pattern, as exhibited at the Liverpool meeting of the Royal Agricultural Society in 1877, with plain leather bucket. The piston-rods of the pump, of the two cylinders, and also of the air-pump condenser, are all in one straight line, though not of the same diameter, being connected by cotters. One rod also serves to actuate the slide-valves of the high and low-pressure cylinders, the valve-chest of the former being mounted on a distance-piece for this purpose. The diameter of the high-pressure cylinder is 21 inches, and of the low 36 inches, the stroke being 4 feet. Steam is admitted into the space afforded by the distance-piece between the high-pressure cylinder and its valve-chest, and makes its way alternately into either end of the cylinder. As soon as the piston, in its course towards the right or left-hand end, passes the orifice of the small pipe, entering a little below the centre line, the steam, still at nearly boiler pressure, passes through this pipe and round the spindle of the small valve, which, by the pressure on its under surface, is lifted off its conical seat, thus allowing the high-pressure steam, which has passed through a small hole drilled in the slide-valve carrier, to exhaust in the space at the back of the main-piston, and thence into the exhaust-passage. The equilibrium of the slide-valve carrier is thus destroyed, as on the other side of it there is live steam in communication with the boiler, so that it is brought over, carrying with it the slide-valve. The small lifting-valve closes again, facilitated by the pressure of the spiral spring with which it is fitted, in conjunction with live steam admitted on the top from the end, and immediately the valve is shifted over. A lever is provided, worked from the outside, for bringing the valve over, in case it should happen to have stopped immediately over the port. The steam exhausts from either end of the high-pressure cylinder into the opposite end of the low-pressure cylinder, the slide-valve of which is attached to the rod of the slide-valve carrier above mentioned, and is therefore actuated by it. The pistons are of the ordinary kind, fitted with cast-iron rings of the locomotive type, and both pistons are attached to one steel rod; the high-pressure piston being screwed on firmly, and the low-pressure one being fastened by a nut and cotter. The air-pump is combined with the condenser, and fitted with rubber disc-valves, gunmetal seats, gunmetal working-barrel, gunmetal cased bucket, and gunmetal pump-rod, the latter being cottered to the low-pressure end of the piston-rod. The condenser is also provided with a suitable injection-valve and spray-pipe. This firm has just supplied a pumping engine of this description for the Chesterfield and Boythorpe Colliery Company. Both the design and workmanship of these pumping engines reflect great credit on the Soho firm, and we are pleased to learn they have given every satisfaction.' (Extract from a newspaper report, May 31, 1878, and revised by the makers.)

The engine supplied to the Chesterfield and Boythorpe Colliery Co., referred to in the foregoing account, is of the following dimensions:—21-inch high-pressure and 36-inch low-pressure cylinders, and a 12-inch double-plunger pump, with a 4-feet stroke. In respect to the working of this engine the manager of the colliery reports as follows:—'The pump is going 12 strokes per minute, and works very regularly. We really should not know the pump was going but for the fall of the clacks when it is changing strokes.'

The following letters give some particulars as to the capabilities of these engines and their behaviour in practice:

'The Chesterfield and Boythorpe Colliery Company, Limited,
'Registered Office, Boythorpe, near Chesterfield.
'*October* 1st, 1879.

'$\frac{21}{36}$" × 12" × 48" *Double-ram Compound Condensing Steam-Pumping Engines.* (Supplied in January 1878.)

'Messrs. Tangye Brothers,—

'Gentlemen,—Referring to the above, which we have now had working continuously night and day for the last twelve months, we are glad to say that it is giving us every satisfaction. It is fixed about 400 feet below the surface, the steam being taken down to it at a pressure of 45 lbs. per square inch. We can work the pump without any difficulty at 28 strokes per minute = 224 feet piston-speed. The pumping power is enormous. The vacuum in the condenser is from 11½ to 13 lbs. The pump is easily started, and works well and regularly, the amount of steam taken being much less than we anticipated. We consider the economy in working very satisfactory indeed. The desire for power and economy at the present day will certainly bring this pump into great requisition. ‘Yours truly,

(Signed) ‘M. STRAW, Manager.

‘Newcastle and Gateshead Water Company,
‘Newcastle-on-Tyne, *October* 20, 1879.

‘$\frac{21}{36}$″ × 10″ × 48″ *Compound Condensing Steam-Pumping Engines.*

‘Messrs. Tangye Brothers,—

‘Gentlemen,—In reply to your inquiry as to the efficiency of the two pairs of compound condensing engines recently erected by you for this company at our Gateshead Pumping Station, I have great pleasure in informing you that they have far surpassed my expectations, being capable of pumping 50 per cent. more water than the quantity contracted for; and by a series of experiments I find they work as economically as any other engine of the compound type, and will compare favourably with any other class of pumping engine. By the simplicity of their arrangement and superior workmanship they require very little attendance and repairs, and the pumps are quite noiseless. A short time ago I had them tried upon air by suddenly shutting off the column, and found they did not run away, thus showing the perfect controlling or governing power of the Floyd's Improved Steam-moved Reversing Valve. I will thank you to forward the other two pairs you have in hand for our Benwell Pumping Station.

‘Yours respectfully’

(Signed) ‘JOHN R. FOSTER (Engineer).’

According to the makers' list of sizes, &c., this engine is made of any size, from 8-inch and 14-inch high and low-pressure cylinders, and 4-inch water-cylinder, and 2-feet stroke, to 30-inch and 52-inch high and low-pressure cylinders, and 14-inch water-cylinder, and 4-feet stroke. The former is capable, with 40 lbs. of steam per square inch, and fitted with an air-pump condenser, of raising about 3,900 gallons per hour 600 feet, and the latter about 47,950 gallons per hour 689 feet. For a list of the sizes and capacities of this engine see page 132.

II.

VERTICAL PUMPING ENGINES.

A. *ROTARY VERTICAL ENGINES.*

THESE engines are not so extensively adopted for the various purposes to which steam-pumps are applied as the horizontal pumps, and are not made in such large sizes. Their use appears to be generally confined to the draining of wells, the pumping out of flooded shafts, or shafts in the course of being sunk, to feeding boilers, and to factories for different purposes. They occupy less room and require much less space laterally than the other forms of pumps. They have been made to pump 1,200 feet vertically in one lift, and may be made to force to greater heights, and are available for all the purposes to which the simple horizontal engines are applied. The most economical steam-pumps are those which are worked on the crank and connecting-rod principle regulated by a fly-wheel, as they permit to a small degree the expansive use of steam; and the stroke being limited by the crank, and the slide-valve operated by an eccentric in the ordinary way, they are more reliable and certain in their action, no damage being possible from the loss of the load in the pump column, as it is impossible for the engine to overrun its stroke, and the piston can be worked much nearer the covers than is safe in non-rotary pumps. The pump is generally of the plunger type, the plunger being packed by an outside stuffing-box, by which the wear is reduced to a minimum. There is also less wear with the steam-piston than is the case with horizontal pumps, and they are more compact than the latter. We are not aware that the principle of compound steam-cylinders has been adopted in these engines, but engines of this form are made with two steam-cylinders, working two ram-pumps, with a fly-wheel common to both. Of course compound cylinders could be easily adopted, but we question whether it would be judicious for such comparatively small pumps. Much of the portability and compactness would be lost, although for large pumps the economy of the compound principle would compensate to a great extent for these drawbacks.

'The rotary motion and fly-wheel give three distinct advantages to steam-pumping machinery not easily to be obtained by any other mechanical contrivance. The first of these is that a perfectly sure and mechanical action of an admission and exhaust-valve is obtained, which can always be relied on at any speed or velocity, and may be rigidly timed to cut off the steam admission, and to open and close the steam-exhaust at any relative time to that of the crank that may be desired. The second is that, by means of a heavy rotating fly-wheel always moving in the same direction, the variable power of steam—first admitted at boiler pressure and then used at a constantly decreasing pressure during expansion,—may be absorbed and equalised so as to overcome, without jerking, or apparent inequality of speed, the constant resisting pressure of the head of water. The third is that the reciprocating motion of a plunger or piston, derived from the constant rotative speed of a crank, varies between no motion at all at either end of the stroke, and a constantly increasing velocity up to maximum speed at the middle of the stroke. This is particularly a motion suited to set a heavy column of water into motion without shock, and to bring it gradually to rest before reversal of the stroke.'*

In some forms of the non-rotative pumps there is the liability of the engine getting 'centred' or coming to a stand-still if it should be worked very slowly. The momentum of the moving mass is sufficient at ordinary speeds to carry the piston beyond the point where at a slow speed it would stop. At a slow speed there is no momentum to overcome the resistance of the water-load after the steam-port has been closed, and the engine will become stationary at either end of its stroke.

* *Direct-Acting Steam-Pumping Machinery*, by Mr. J. C. Fell, of London. A paper read before the South Staffordshire and East Worcestershire Inst., Feb. 19, 1877.

The advocates of the non-rotary engines claim greater simplicity of parts than is to be found in the rotary engines, but this is not true of some of the former. There is no doubt that the rotary pumps as a class are more solidly constructed than the non-rotary pumps, and the details are of a more durable character. It is a most important point in a mining pump to avoid delicate details, and more especially when they are not easy of access for examination or repair.

In the following classification, which is a recapitulation of that which appears at page 8, it will be seen that the pumps are arranged according to the description of the pump part. A classification founded upon the character of the steam part would not be sufficiently definite, as the engines are all simple, the only difference being that some engines have one cylinder whilst others have two cylinders.

a. One steam-cylinder and one single-acting ram.
b. One steam-cylinder and two single-acting rams.
c. Two steam-cylinders and two single-acting rams.
d. One steam-cylinder and one double-acting ram.
e. Two steam-cylinders and two double-acting rams.
f. One steam-cylinder, one double-acting piston, and D slide-valve pump
g. One steam-cylinder and one single-acting ram, and single-acting piston.

A. SIMPLE STEAM-PUMPS.

THESE engines are often known as 'donkey-engines,' and are applied to a great variety of purposes. The smaller kinds are extensively used in the navy for feeding boilers and pumping out water ballast and bilges, whilst in chemical-works, tanneries, gas-works, breweries, &c., they find an unlimited sphere of usefulness, although perhaps not so largely employed as the non-rotary horizontal pumps. They are usually restricted to a small size, but for mining purposes are made sometimes as large as 20-inch steam-cylinders and from 15-inch to 20-inch rams.

In the simplest varieties of these pumps the ram is single-acting, forcing only at the down stroke. Other forms have a pair of single-acting rams, and in others, by the addition of another ram-case and suction and delivery-valves, the ram has a double-action, forcing both at the down and up strokes. Some pumps have two such rams and are known as quadruple-acting pumps. Two classes of pumps widely differing from any of the foregoing remain to be noticed. In one of these instead of a ram a piston is used, and a D slide-valve instead of the lifting-valves. In the other class the pump consists of a single-acting ram combined with a single-acting piston.

Sizes and Speed.

These pumps are made in sizes varying from 4-inch to 20-inch steam-cylinders, and from 1¼-inch to 15-inch pumps, and from 3-inch to 15-inch strokes, but the principle is of course not limited to these sizes. The smaller pumps are useful more especially for feeding boilers and for other uses in factories, steamboats, &c. As to speed, these engines may be driven as fast as non-rotary engines, and are safer at high speeds, owing to the stroke being limited by a fly-wheel, and it being accordingly impossible for the engine to overrun its stroke. The following particulars refer to Messrs. May and Mountain's pump:

Diameter of steam-cylinder	Diameter of plunger	Length of stroke	Gallons delivered per hour, at 70 revolutions per minute	Gallons delivered per hour, at 60 revolutions per minute	Gallons delivered per hour, at 50 revolutions per minute	Piston speed at 70 revolutions	Piston speed at 60 revolutions	Piston speed at 50 revolutions	Diameter of steam-pipe	Diameter of exhaust-pipe	Diameter of suction-pipe	Diameter of delivery-pipe
in	in.	in.				ft.	ft.	ft.	in.	in.	in.	in.
10	5	12	6420	5500	4590	140	120	100	1½	2	5	4
8	5	12	6420	5500	4590	140	120	100	1¼	1¾	5	4
8	4	9	3080	2630	2200	105	90	75	1¼	1¾	4	3
6	4	9	3080	2630	2200	105	90	75	1¼	1¾	4	3
5	3¼	7½	1690	1550	1210	87½	75	62½	1¼	1¾	3½	2¼
4	2½	6	800	680	630	70	60	50	1	1½	3	2
4	2¼	6	650	550	460	70	60	50	1	1½	3	2
3	2	4½	380	330	270	52½	45	37½	¾	1¼	2½	1½
3	1½	4½	216	185	154	52½	45	37½	¾	1¼	2½	1½

Capabilities.

The best forms of vertical pumps of the rotary class, such as those of Mr. John Cameron, Messrs. Ashworth Brothers, and similar pumps of other makers, are suitable for the deepest and most extensive drainage operations performed by any other class of simple steam-pumps. In one large colliery district in South Wales which the author visited, pumps of this class were reputed to be the favourites.

Two years ago Mr. Cameron supplied to Jones's Navigation Colliery, Rhondda Valley, South Wales, a 3-inch double ram-pump, with two 11-inch steam-cylinders, to force to 1,125 feet. A great number of Mr. Cameron's pumps are in use in deep pits, forcing up to 1,200 feet vertically.

Messrs. Ashworth Brothers are making for a Welsh colliery an extra strong 'Simplex' quadruple-acting ram-pump, with two 25-inch steam-cylinders, two 4¾-inch rams and 9-inch stroke, to force water 1,200 feet in a single vertical lift, and Messrs. Hulme and Lund have made for a Lancashire colliery a small pump to force to 1,140 feet in one lift.

These pumps, owing to the stroke being limited by a crank, can be worked at their highest speed with safety, and through the regulating and compensating medium of a heavy fly-wheel there is great regularity in the strokes. As the steam-valve is operated by the certain and unerring movements of an eccentric on the crank-shaft, no erratic action of the valve can ensue, and the steam may be cut off at any portion of the stroke with the greatest exactitude; a saving of steam is thus effected.

In low galleries, small levels, and in any workings which have very little vertical space, engines of the horizontal type are more suitable, but this is a circumstance that cannot militate much against the adoption of vertical engines in mines and collieries for general light drainage.

Cost.

By comparing the prices given opposite with those of the non-rotary pumps, it will be found that the single-ram double-acting pumps are slightly more costly than non-rotary pumps of a similar size, and that the double-acting ram-pumps with two rams cost from about 30 to 65 per cent. more than corresponding non-rotary pumps. The following are the prices of a variety of ram-pumps specially designed and adapted for underground drainage. About 10 per cent. extra is charged for governors, &c. for small engines, and about 5 per cent. for large engines.

The second table gives the prices of engines of the piston and slide-valve pump class, such as those of Messrs. Carrick and Wardale and William Turner.

Single-acting pumps					Double-acting pumps										Quadruple-acting pumps				
One ram					One ram					Two rams					Two rams				
Diameter of steam-cylinder	Diameter of ram	Length of stroke	Delivery per hour	Price	Diameter of steam-cylinder	Diameter of ram	Length of stroke	Delivery per hour	Price	Diameter of steam-cylinder	Diameter of ram	Length of stroke	Delivery per hour	Price	Diameter of steam-cylinder	Diameter of ram	Length of stroke	Delivery per hour	Price
in.	in.	in.	gallons	£	in.	in.	in.	gallons	£	in.	in.	in	gallons	£	in.	in.	in	gallons	£
4	2	3	240	16	4	2	3	446	22	4	2	3	480	33	4	2	3	892	40
5	2½	4	400	20	5	2½	4	714	26	5	2½	4	800	38	5	2½	4	1,428	45
6	3	5	650	24	6	3	5	1,030	32	6	3	5	1,200	46	6	3	5	2,060	55
6	3½	5	900	26	6	3½	5	1,440	35	6	3½	5	1,800	49	6	3½	5	2,880	60
7	4	6	1,200	32	7	4	6	1,863	43	7	4	6	2,400	56	7	4	6	3,726	75
8	4½	6	1,600	34½	8	4½	6	2,385	48	8	4½	6	3,200	61	8	4½	6	4,770	85
8	5	6	2,000	37	8	5	6	2,965	54	8	5	6	4,000	66	8	5	6	5,930	95
10	6	8	3,200	48	10	6	8	4,268	70	10	6	8	6,400	86	10	6	8	8,536	125
—	—	—	—	—	12	6	8	4,268	75	12	7	9	9,500	110	12	6	8	8,536	135
—	—	—	—	—	12	7	9	5,600	90	12	8	10	12,800	140	12	7	9	11,200	170
—	—	—	—	—	14	8	10	7,650	115	14	10	12½	22,800	220	14	8	10	15,300	210
—	—	—	—	—	16	10	12½	11,760	155	16	12	15	36,000	300	16	10	12½	23,520	280
—	—	—	—	—	—	—	—	—	—	—	—	—	—	—	20	12	15	34,000	380
—	—	—	—	—	—	—	—	—	—	—	—	—	—	—	20	15	15	—	—

Diameter of steam-cyl.	Diameter of pump	Length of stroke	Delivery per hour	Price			Extra for brass lined pump and brass-coated rods		
in.	in.	in.	gallons	£	s.	d.	£	s.	d.
6	4	6	1,800	45	0	0	2	0	0
8	5	8	4,500	55	0	0	2	10	0
10	6	10	9,000	82	10	0	3	0	0
12	8	12	20,000	115	0	0	4	0	0

SINGLE STEAM-CYLINDER, EXPANSIVE, NON-CONDENSING ENGINES, WORKING PUMP BY CONNECTING-ROD AND CRANK, REGULATED BY A FLY-WHEEL.

A. ONE STEAM-CYLINDER AND ONE SINGLE-ACTING RAM.

CAMERON'S SINGLE-ACTING RAM PUMP (1ST FORM).

This admirable steam-pump was invented by Mr. John Cameron, of the Oldfield Road Iron Works, Salford, Manchester, in 1853, and is one of the oldest and best known engines of the class in use, and many thousands have been made. Since its introduction it has been gradually improved; the engine shown in the annexed woodcut, fig. 108, is one of the latest developments of the principle, and is patented.

This engine consists of a steam-cylinder, with an ordinary three-ported slide-valve worked direct from an eccentric on the crank-shaft, and mounted on two cast-iron columns, which serve as delivery air-vessels to the pump. It will be observed, on reference to fig. 108, that the cylinder-bottom, the columns referred to, the valve-chambers, and the ram-case are one casting, and form their own bed-plate. The arrangement of the water part is very simple. Upon the end of the ram, and forming part of the same casting, is a triangle, to the top end of which a piston-rod is cottered, and a little below this is the wrist-pin for the small end of the connecting-rod. The fly-wheel is mounted upon a shaft supported upon brackets cast to the columns. The suction and delivery-valve-chambers are arranged directly under the air-vessel, and from their position are easily accessible for examination and repair. The suction and delivery-branches are at the lower part of the pump, as shown in the illustration. We might add that the arrangement is a very compact one, and from the substantial character of the parts is not likely soon to become disordered.

The account given below, which has been supplied to the author, will be found useful:

'The steam-engine part is the simplest possible. The cylinder is supported by the two delivery air-vessel columns of the pump. The admission and exit of the steam is regu-

FIG. 108.

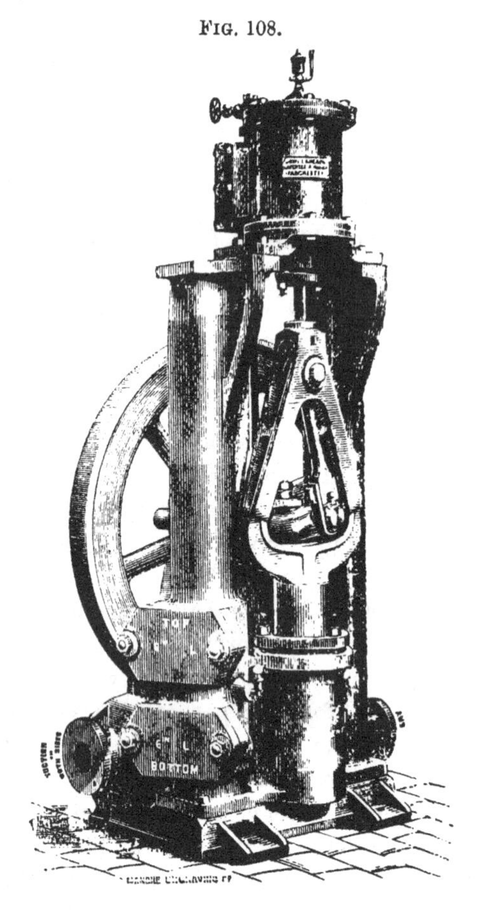

lated by the ordinary simple three-port slide-valve, worked direct from an eccentric on the crank-shaft below, and which revolves in gun-metal steps, and carries the fly-wheel, which is always turned and runs in balance. The piston-rod is cottered into the head of the pump-ram casting, and a little below is the wrist-pin for the small end of the connecting-rod. The lower part is widened out into a triangular shape, to allow room for the lower end of the connecting-rod to swing inside at the crank end. To the bottom of the triangle the pump-ram is cast, and works through a gland and packing, which is supported by a loose brass collar-bush at the bottom of the stuffing-box. The ram descends into the chamber below, and communicates direct, by a short passage, with the pump-valves. The top or delivery-valve allows the water to pass direct into the air-vessel above, whence it flows into the second air-vessel column. where its flow is further equalised before its final exit. The suction-valve communicates direct into the vacuum-vessel of the pump, whence, on the rise of the ram, it gets its intermittent supply, due to its displacement. The water taken from the vacuum-vessel is replenished by a constant flow from the well, caused by the atmospheric pressure on the water, and thus the constant flow or momentum of the water is maintained, and shocks to the suction-pipe avoided. The valves are of the ordinary mushroom vertical-acting class, with centre guide spindles, and are usually the same diameter as the pump. Although guards are fitted to the valve-doors, they are seldom, if ever, required, for the reason that the lift of the suction-valve in particular should be in accordance with the requirements of the displacement. The water being forced through this valve by the atmospheric pressure only, its opening is hindered by its coming in contact with a stop, the opening is therefore too small, and consequently obstructs the passage of the water in sufficient quantity to replenish the displacement, and the pump-chamber is only partially filled. In such a case the descent of the pump-ram causes a shock more or less severe, depending on the pressure.'

At Jones's Navigation Colliery, Rhondda Valley, South Wales, which the author visited in June 1877, there were three of these pumps at work in the pit, one forcing about 405 feet, and two forcing about 180 feet. One pump had a 13-inch steam-cylinder, and the others 12-inch steam-cylinders; the rams were 5½ inches, and the stroke 8 inches.

the delivery-pipes were 4 inches in diameter, and the rising-main 12 inches. The average number of strokes was 60 per minute. The steam-pipes were 2 inches, and the steam was taken down from the surface. A smaller Cameron's pump (about 5-inch ram and 6-inch stroke) was in use for feeding the boilers of the winding-engines. In the Massinethmoor pit of the same colliery the 'Special' and 'Universal' pumps were in use. The 'Special' and 'Cameron' were working upon the same stage in the pit, forcing water 90 feet, to the 'Universal' pump, which was forcing 54 feet.

About two years ago a 3-inch Cameron's double ram-pump with two 11-inch steam-cylinders was supplied to the same colliery to force to 1125 feet.

Mr. Cameron's pumps are highly spoken of in the Rhondda Valley, and are generally preferred to other pumps. They give very little trouble; the valves are very durable, and are easily repaired. The valves are spindle-valves of the type sometimes known as mushroom valves. They are of tough gun-metal, as are also the seats. In the colliery above mentioned there had been no new valves during the time the engineer had been in the colliery—a period of about 16 months. The pumps had been at work about three years. Mr. Cameron recommends that in fixing the pump the suction-pipes should be well tested by water-pressure, and the joints made perfectly tight. Air-leaks in the suction-pipe are generally attributed to a defect in the pump, and are very difficult to discover when the pipes are laid. If the pump has to draw above a few feet vertically, a roomy stop-valve should be fitted in the suction-pipe, near the water. Mr. Cameron also makes double-ram pumps with one or two steam-cylinders. The advertised sizes of this pump vary from a 2-inch to an 18-inch ram; and a 3-inch to an 18-inch stroke. The larger sizes are excellent fire-engines.

Mr. Cameron claims that his pumps, with their crank-limited strokes, are the most reliable, durable, and simple pumps known, and that they require less skilled attention than any other kind of pump.

ASHWORTH BROTHERS' SINGLE-ACTING RAM-PUMP (1ST FORM).

Messrs. Ashworth Brothers, of the Moss Brook Foundry, Manchester, are the inventors and sole makers of the following varieties of vertical ram-pumps, namely : single-acting, with one ram ; double-acting, with one ram ; double-acting, with two rams ; and quadruple-acting, with two rams.

This pump is advertised in sizes varying from 4-inch to 8½-inch steam-cylinders, and 2-inch to 6-inch rams. The stroke of the sizes named is 3 inches and 8 inches respectively, and the hourly discharge of water is 280 and 2,500 gallons.

The bed-plate is hollow, and acts as an air-vessel and a reservoir for water, and the columns which support the steam-cylinder are air-vessels for the delivery-pipe, and the piston and valve-rods are of steel, working through brass glands and bushes. The valves are of brass.

From 7-inch rams inclusive, and upwards, the valves of these engines have the same area as the ram, and all the water-ways are made larger in proportion. This important improvement enables users to run the engine at a greater speed, and consequently deliver more water, and it also promotes the durability of the valves.

THE 'MANCHESTER' SINGLE-ACTING RAM-PUMP (1ST FORM).

It will be observed from fig. 109 that this pump differs from those previously noticed in having the crank-shaft under the pump, and the fly-wheel consequently fixed almost at the bottom of the engine. The vibration-loop takes hold of the piston-rod and ram in a different manner from the other kinds of engines, and being so long the side-thrust due to the crank is inconsiderable. The advantage of the fly-wheel so low down is that the engine is steadier, and even very large engines require no foundation-bolts.

This arrangement is less compact than that of the other engines of this class, and in shafts where space is very limited it is not so suitable a machine. We suppose it is

Fig. 109.

Fig. 110

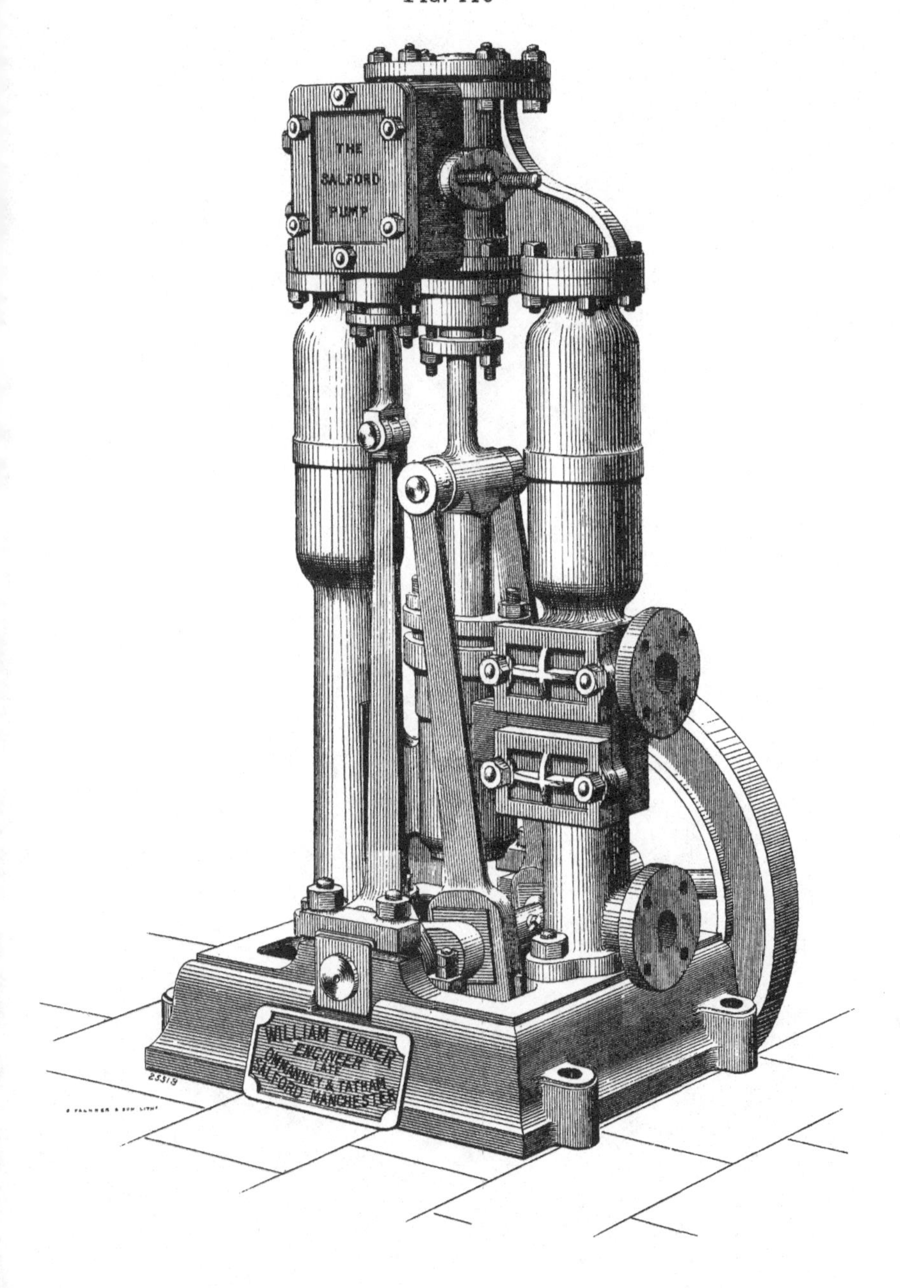
THE
SALFORD
PUMP
WILLIAM TURNER
ENGINEER
LATE
OMMANNEY & TATHAM
SALFORD MANCHESTER
25318

not intended for use underground in mines, but is specially designed for feeding boilers.

It is made, according to the lists of the makers, Messrs. Frank Pearn, Wells & Co., West Gorton, Manchester, in sizes varying from $1\frac{1}{4}$ to 6-inch rams, and $2\frac{1}{2}$ to 10-inch strokes, delivering from 130 to 3,400 gallons per hour.

THE 'SALFORD' SINGLE-ACTING RAM-PUMP.

This engine is arranged with an air-vessel both on the suction and the delivery side. The connecting-rods, it will be seen, are long, and the side-thrust accordingly little. The arrangement, which is a very neat one, is in some respects similar to the 'Manchester' pump. The two pillars are air-vessels; the suction air-vessel is shown at the left-hand side of the accompanying fig. 110 and the delivery air-vessel at the right-hand side over the valve-chambers. The connection is made below the crank-shaft by a pipe cast in the bed. The connecting-rods are perfectly straight, and with large bearings. One connecting-rod is sufficient, but for uniformity, strength, and steadiness Mr. Turner prefers to use two.

The pump valves are of gun-metal, of the mushroom type, the central bearing being a spindle, and both these and the steam-valves are easily accessible for adjustment and repair. It is made in sizes varying from a $2\frac{1}{2}$-inch to a 9-inch steam-cylinder and a $1\frac{1}{4}$-inch to a 6-inch ram, the stroke varying from $2\frac{1}{2}$ to 8 inches, and the delivery from 100 to 3,200 gallons per hour. Mr. Turner states that he can make this type up to 18-inch cylinders and 8-inch pumps. It can be had double-acting.

Mr. William Turner, of Salford, is the maker of this engine.

HULME AND LUND'S SINGLE-ACTING RAM-PUMP (1ST FORM).

Messrs. Hulme and Lund, of Salford, are makers of single-acting ram-pumps of similar construction to those already mentioned. By the application of a patent air-chamber above the suction-pipe, on the base-plate, the agitation in the suction-pipe is avoided, and the water has a more constant flow into the pump-barrels, so that they get properly filled at each stroke. The air-chamber also greatly reduces the wear and tear of the water-valves, and being well above the suction-pipe, they cannot get choked up by any collection of sediment, as the passages are cleaned out with every stroke of the pump. This pump is advertised in sizes varying from a 2-inch to a 6-inch ram, with a 3-inch to 8-inch stroke, but of course it can be produced in any size that may be ordered.

B. ONE STEAM-CYLINDER AND TWO SINGLE-ACTING RAMS.

CAMERON'S DOUBLE RAM-PUMP (WITH ONE STEAM-CYLINDER; 2ND FORM).

The principle of action in this pump is precisely the same as that of the single ram-pump. There are two single-acting rams, one connected directly underneath the steam-cylinder, from which it obtains its motion in the same manner as in the pump previously described, the second ram receiving its motion from the same crank-shaft. To the top of the triangle of the second ram is a steel rod working through a guide at the top of the two columns.

The cranks are set opposite to each other, consequently one pump is delivering its water at the moment the other is receiving it. By this means the pressure on the piston is equalised at each side, and a continuous discharge is maintained in the rising-main.

This form of engine is specially useful when a large quantity of water has to be raised a short distance. It would do the same work as the pump with two cylinders at less first cost, and would compare favourably with it as regards the quantity of steam used. This engine is often used in gas and chemical works for pumping different liquids at the same time. If an accident happened to one pump, the other could be kept at work.

C. TWO STEAM-CYLINDERS AND TWO SINGLE-ACTING RAMS.

CAMERON'S PATENT DOUBLE RAM-PUMP (WITH TWO STEAM-CYLINDERS; 3RD FORM).

This pump, fig. 111, is a new patent, and differs from the one-cylinder double ram-pump in having another cylinder,

FIG. 111.

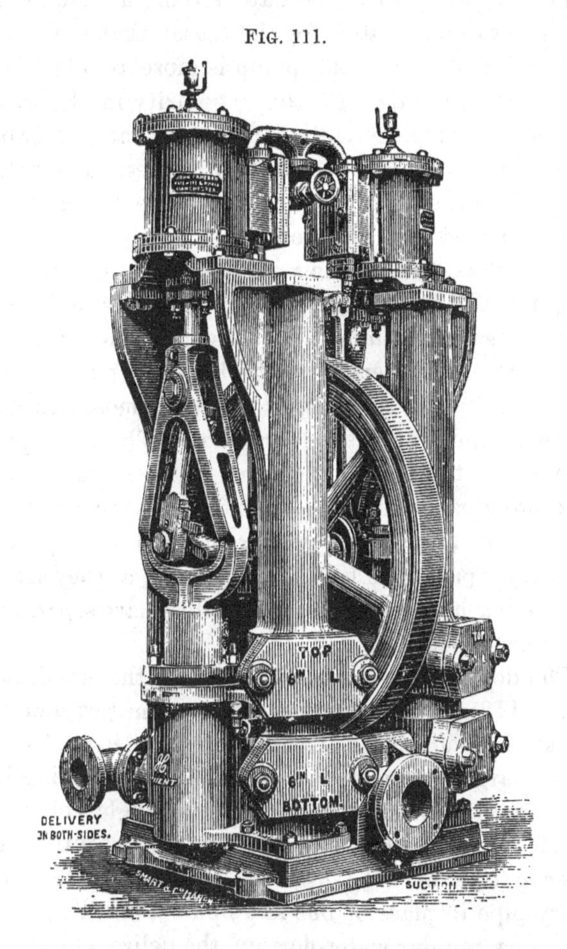

and consequently is more suitable for forcing against great pressures. If it should be decided to adopt an engine of this class, preference should be given to this over the last engine described.

The following particulars have been communicated to the author:

'So much for the single-acting lift and force-pump (1st form), which is preferred for moderate pressures chiefly on account of its simplicity. But for high pressures the double ram-pump, with alternate forcing action, is a better method of forcing water, for the reason that the flow of the water into and out of the pump is more regular, therefore the air-vessels produce greater regularity in the water-flow. Two kinds of this description of double-ram, or two single-acting ram-pumps coupled together by steam and water-pipes and by the crank-shaft, are made. The annexed woodcut (fig. 111) shows the class of steam-pump chiefly used for feeding boilers, and forcing water up to 100 lbs. pressure per inch. The two pumps are bolted upon a bed-plate; the suction-passage is common to both, and the cranks are set opposite to each other, so that when one pump draws the other forces. The pumping action is almost noiseless, and the flow of the water is very regular. The arrangement is compact, and takes but little room. This class is worked by one or two steam-cylinders, depending on the pressure; two cylinders, one above each pump-ram, reduce the pressure on the working parts one half, each cylinder, as they are double-acting, doing half the work. The valves have separate doors, and are easy of access.

'The double ram-pump, as shown by the woodcut opposite (fig. 112), is used for the largest quantities and for the heaviest pressures. It is used for water-works, water-lifts, for large iron and chemical works, for feeding boilers, and for forcing water in coal-pits in a single column up to 1,500 feet. As little standing room is required, several smaller pumps may be used and connected to the same delivery-pipe in place of one large pump, and thereby maintain a very regular water-flow up the delivery-pipe column, and in case any of them need repacking or repair, its stoppage causes but little inconvenience. This pump is differently constructed from the two before described, inasmuch as there are no water-passages in the bed-plate—which in this pump is only a frame or plate to which the pump columns are bolted and steadily pinned, so that they are easily taken to pieces to get down pits, and as easily put together

FIG. 112.

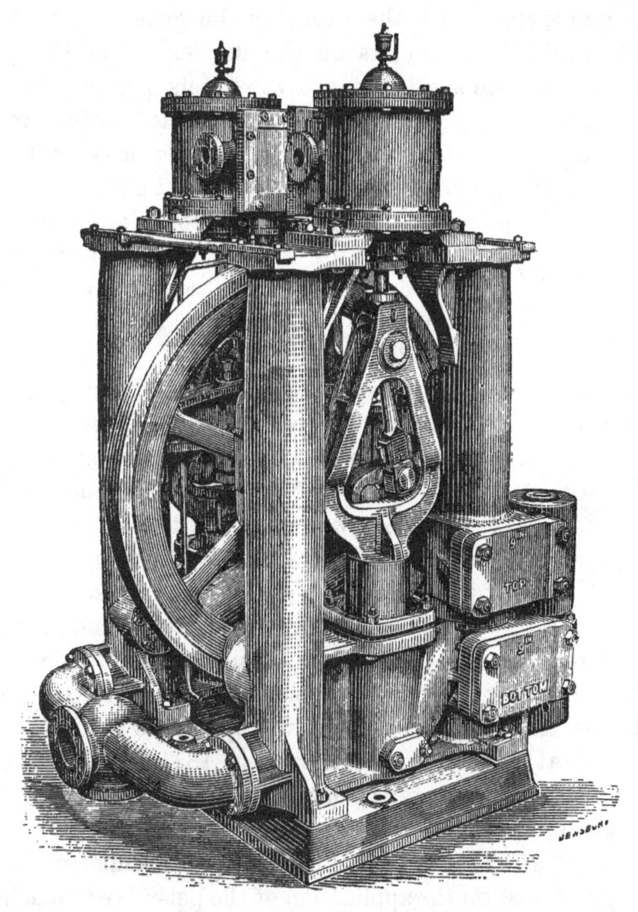

again by an ordinary workman. The largest made for forcing up a coal-pit shaft had 18-inch double-rams with 18-inch stroke, and its weight was about 27 tons. It enabled the owners to stop a large beam-lift engine, and it has worked several years satisfactorily, forcing the water up a single delivery column. In case of the exhaust steam being objectionable in colliery or mine workings it can, in most cases, be coupled with the suction-water and condensed, and vacuum power added to the pistons by the partial removal of the pressure of the atmosphere. A governor is sometimes applied, which is driven by bevel wheels, one on the crank-shaft and the other on the governor-spindle, so that in case a pipe bursts on the delivery side, the pump draws air on the suction side, a valve gets gagged, or from any other cause the pump is relieved of its burden, or the pressure of the steam varies, no slipping may occur, the action of the governor keeping the pump at a nearly uniform speed. The pump may be worked with considerable economy of steam by the addition of arrangements for cutting it off and working expansively, and also by attaching an air-pump and condenser.

FIG. 113.

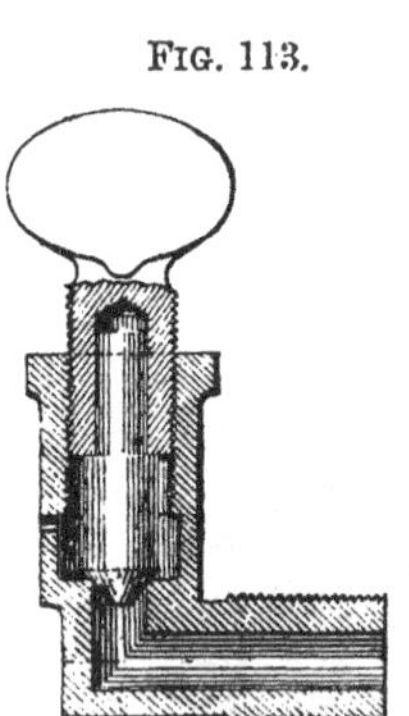

'A small pet-valve (as annexed, fig. 113), is inserted under the delivery-valve or highest part of the pump to allow egress of air or water therefrom, and so indicate its working condition. This valve is the common vertical lift-valve, and has a thumb-screw over it; on unscrewing the latter the valve acts the same as the delivery-valve of the pump, closing during the suction-stroke, and preventing the admission of air which would vitiate the action of the pump. This valve was applied to vertical-acting plungers at a water-works in the year 1841 by Mr. Cameron. The pumps drew water from the river Elbe, and worked all right when the tide was in, but at low water they would not draw at all, on account of the air accumulating in the plunger-cases; on the application of the pet-valve to discharge the air the reliable action of the pumps was secured. Mr. Cameron believes this was the first application of such a pet-valve to pumps. Previous to this, a common small cock was, and is still occasionally used for the purpose. The pet-valve is in almost universal use wherever known, and is a little but useful adjunct to all kinds of ram or piston-pumps.

'It will be observed on reference to the woodcuts that all the working parts are easy of access and capable of adjustment; the pumps are made throughout from a repair point of view, every important part being made separately to gauge and template, so that in case of accident the part affected can be replaced with facility. The bearing-steps are all square-bottomed, so as to "line up" easily when worn. Common head or cotter-bolts are used in preference to screws or studs for all parts that disengage, as the trouble caused by a broken stud or screw in a joint is well known to those who have had to contend with such accidents and with inferior tools.'

Ashworth Brothers' Double-acting Ram-Pump (with two rams; 2nd form).

Fig. 114.

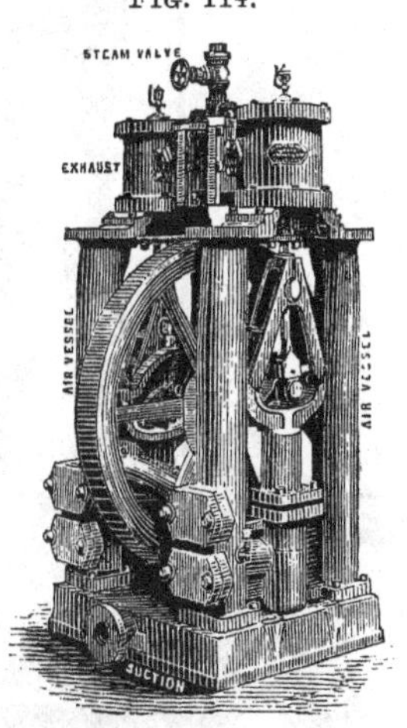

This pump does not differ materially from Mr. Cameron's pump of the same type. It is made in sizes of from 4-inch to 20-inch steam-cylinders and 2-inch to 15-inch rams, according to the makers' list. The accompanying illustration, fig. 114, is an elevation.

The 'Manchester' Double-acting Ram-Pump (with two rams; 2nd form).

This engine is simply an extension of the previously noticed engine of Messrs. Frank Pearn, Wells and Co. Fig. 115 is a perspective elevation.

Fig. 115.

HULME AND LUND'S DOUBLE-ACTING RAM-PUMP (WITH TWO RAMS; 2ND FORM).

This pump has two single-acting rams, and is of similar construction to the preceding pumps. It is fitted with a patent suction air-vessel, reference to which has already been made.

Pumps of this kind are used in pits in many parts of the country, the heights of delivery varying from 1,140 feet downwards, the average height being about 700 or 800 feet. Messrs. David Brownlow and Co., Gurnwood Park Colliery, St. Helen's, Lancashire, have a small pump of this class, with 3-inch rams and 6-inch stroke, raising about 1,600 gallons per hour, 1,140 feet in one lift.

D. ONE STEAM-CYLINDER AND ONE DOUBLE-ACTING RAM.

THE 'SIMPLEX' PATENT DOUBLE-ACTING RAM-PUMP (WITH ONE RAM; 3RD FORM).

This pump differs from the single ram-pumps previously described in having an additional ram-case and set of suction and delivery-valves in order to constitute it a double-acting machine. We observed in the other pumps described that double-action was only secured by a combination of two single-acting rams.

FIG. 116.

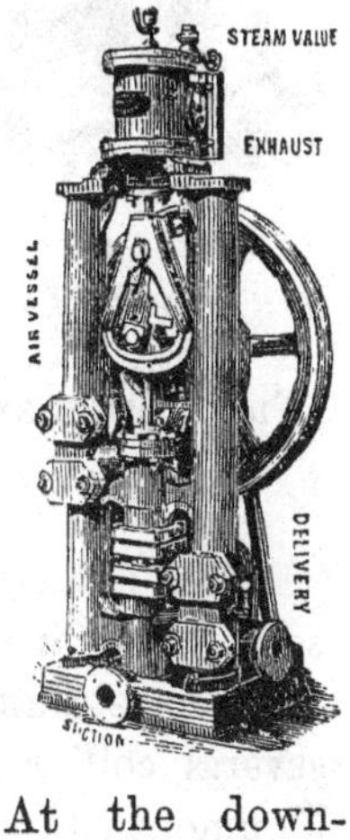

STEAM VALVE
EXHAUST
AIR VESSEL
DELIVERY
SUCTION

The ram-cases are fixed over each other, the ram being attached to a small rod or lifting piece working through a stuffing-box at the top of the upper ram-case. The ram passes through two stuffing-boxes in the position shown.

The two columns supporting the steam-cylinder act as air-vessels, one for each ram-case. The action is the same as that of a double-acting piston-pump. At the down-stroke it delivers from the bottom ram-case, and at the up-stroke it delivers from the upper ram-case. Of course it is a much more durable machine than a piston-pump. Fig. 116 is an elevation. The sizes advertised by the makers range from 4-inch to 16-inch steam-cylinders and from 2-inch to 10-inch rams.

Frank Pearn, Wells and Co.'s Double-acting Ram-Pump (1st form).

Fig. 117.

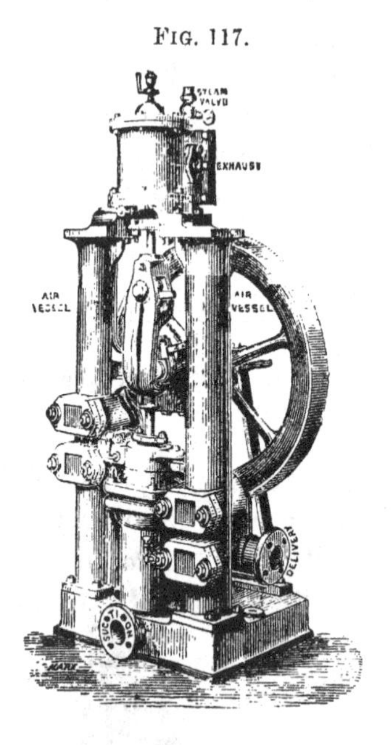

This is a very similar engine to the one just noticed. It is made in sizes varying from a 2-inch to 10-inch ram, a 4-inch to 16-inch steam-cylinder, and a 3-inch to 12½-inch stroke. The accompanying woodcut, fig. 117, is a perspective elevation.

E. TWO STEAM-CYLINDERS AND TWO DOUBLE-ACTING RAMS.

The 'Simplex' Patent Quadruple-acting Ram-Pump (with two rams; 4th form).

This is a combination of two such pumps as that last described, but it has only one fly-wheel and crank-shaft. This is the most elaborate machine of the class, and is in use in several collieries. Messrs. D. and D. Pryce, of Newport, Monmouthshire, have two of these engines at work, each with 8-inch rams. Messrs. The Energlyn Coal Company,

Pwll-y-pant, near Cardiff, have three 8-inch pumps of this kind at work in their collieries; Messrs. The Bettisfield Colliery Company, Bagillt, Staffordshire, have a 5-inch quadruple-acting pump, forcing water more than 900 feet high in one lift, and at numerous other collieries these pumps are at work. Fig. 118 is an elevation of the machine.

FIG. 118.

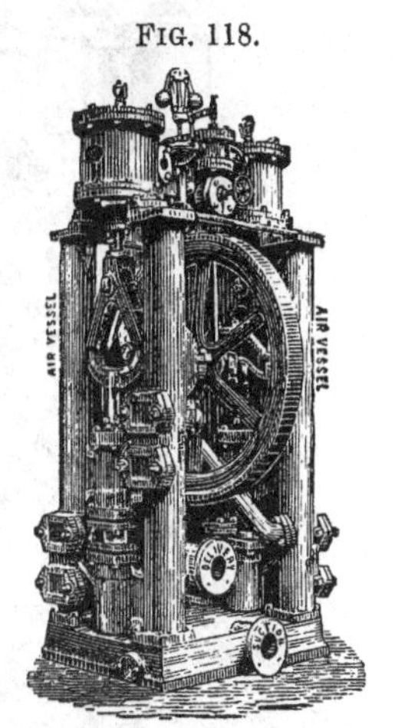

These pumps are being made with steam-cylinders varying from 4 to 25 inches in diameter, and they can be produced in any combination of rams and steam-cylinders, according to the head of water and the steam-pressure at disposal. Messrs. Ashworth Brothers are at present making one of these pumps for Messrs. The Cymmer Coal Co., Porth, near Cardiff, with $4\frac{3}{4}$-inch rams, and 25-inch steam-cylinders, and 9-inch stroke, for forcing water 1,200 feet in one vertical lift. This pump will have all the makers' latest improvements. The columns where the water-valves fit are enlarged, being the size generally used for 7-inch pumps, and the valves are 5 inches bore, being larger than the rams, and have a lift of only $\frac{5}{16}$ of an inch, and are made of gun-metal with very wide faces, to ensure durability. The pump is fitted with four 'priming' attachments (one on each column), so that by turning a handle the pump-barrels may be filled from the delivery, to facilitate the starting of the pump, after 'standing.'

These pumps are largely used in the Lancashire Cotton Mills as fire engines.

FRANK PEARN, WELLS AND CO.'S QUADRUPLE-ACTING DOUBLE RAM-PUMP (2ND FORM).

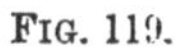

FIG. 119.

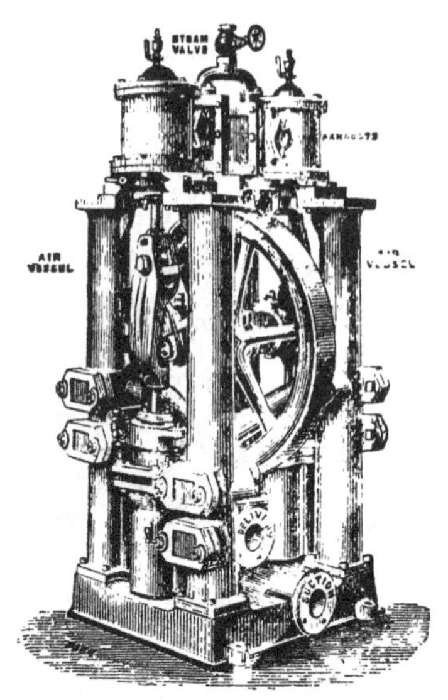

The above firm are makers of a quadruple-acting ram-pump of similar construction to that last mentioned, an illustration (fig. 119) of which will be sufficient.

This engine is made in sizes varying from a 2-inch to 12-inch ram, and a 4-inch to 20-inch steam-cylinder, and from 3-inch to 15-inch stroke.

F. ONE STEAM-CYLINDER AND ONE DOUBLE-ACTING PISTON AND SLIDE-VALVE PUMP.

CARRICK AND WARDALE'S IMPROVED DOUBLE-ACTING PISTON-PUMP, WITH D-SLIDE PUMP-VALVE.

This engine was introduced in the early part of the year 1873. It contains a piston-pump with a slide-valve of the ordinary D type, instead of lifting-valves, as in the other varieties of engines described. Pumps of this class, although very useful for surface-pumping, are not, we think, adapted for mines and collieries where the water is apt to be charged with grit, small fragments of rock, coal dross, &c. They might be used underground if it could always be arranged for them to pump from a 'still pool,' otherwise we could not recommend their adoption. They are specially adapted for use as donkey-pumps for marine purposes; the valve of the pump being a slide instead of the ordinary lifting valves, there is no noise whatever in working, and the thumping of the force-pump cannot annoy the passengers, and they can be run at very high speeds with efficiency.

As will be observed from fig. 120, the pump slide-valve is

Fig. 120.

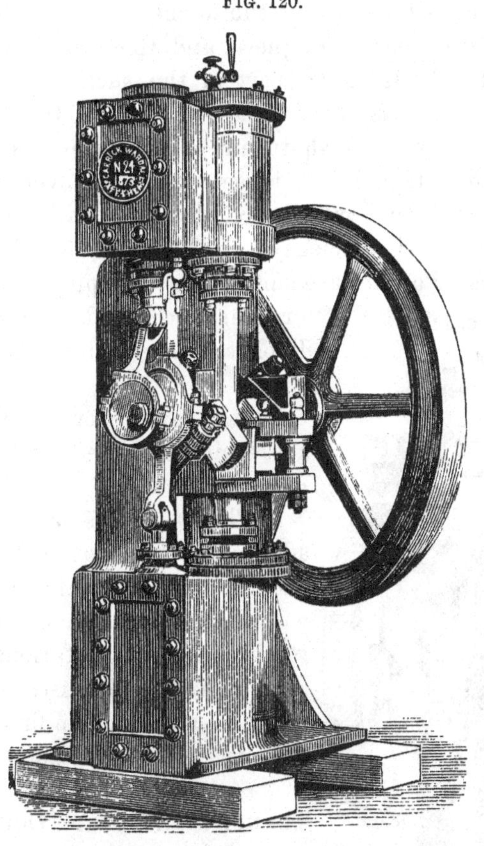

worked by an eccentric on the crank-shaft, as is also the slide-valve of the engine. The action of the suction and delivery being regulated by the movement of the slide-valve, the action is consequently absolutely certain, which is not the case with pumps having lifting-valves.

The makers, Messrs. Carrick and Wardale, Redheugh

Engine Works, Gateshead-on-Tyne, report as follows in their prospectus :

'These pumps have been designed to be simple in arrangement, strong in construction, easy of access to all working parts, not liable to get out of order, efficient and durable.

'They are constructed to work against high-pressures, and are all tested before leaving the works.

'They are cast in one piece, and the standards between pump and cylinder form part of the suction and delivery. The pump-valve is an ordinary D slide, actuated by an eccentric on the crank-shaft, and is of easy access.

'The advantages which these pumps have over those with the common lifting-valves, are, that they will run at very high speeds with efficiency, are noiseless, will draw water from a greater depth, are suitable for pumping either water or semi-fluid liquids, without being liable to choke with pieces of chips, waste, rope, grain, small coals, &c., &c., have fewer working parts, and are easily managed by unskilled men.

FIG. 121.

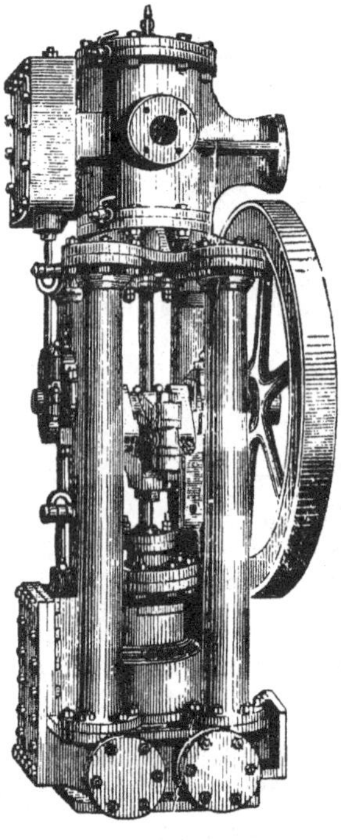

'These pumps are specially adapted for use at collieries, in chemical works, for feeding boilers with either hot or cold water, for railway watering stations, ballast pumps, and other uses on board steamers, and as steam fire-engines.' For prices see table, page 198.

TURNER'S IMPROVED DOUBLE-ACTING PISTON-PUMP, WITH D-SLIDE PUMP-VALVE.

Mr. William Turner, of Salford, is the maker of an engine of this class of somewhat similar construction to that of Messrs. Carrick and Wardale, and of which fig.

121 is an illustration. This engine has been specially designed for marine purposes, as a donkey-pump. Like the engine last described, it is not suitable for mine drainage except to a very limited extent. The pump-barrel is lined with brass, and the pump-piston is coated with the same metal; the pump-valve, valve-rod, and the glands of the same are made entirely of brass, and the valve-face is coated with brass. For prices, see table, page 198. The following is a list of sizes :

Diameter of cylinder	Height over all		Breadth over all		Depth over all		Diameter of ram	Stroke	Steam-pipe	Diameter of steam pipe flange	Diameter of exhaust-pipe	Diameter of exhaust-pipe flange	Diameter of suction and delivery-pipes	Diameter of suction and delivery-pipe flanges
in.	ft.	in.	ft.	in.	ft.	in.	in.	in.	in.	in.	in.	in.	in.	in.
3	3	0	1	6	1	0	1½	4½	¾	3⅛	⅞	3	1	4
3¾	3	2	1	8	1	2	1⅞	5	¾	3¼	1	3¼	1¼	4¼
4½	3	3	1	8	1	5	2¼	4½		3⅜	1¼	3½	1½	5¼
6	4	1	2	0	1	8	3	6	1	3½	1½	3¾	2	5¾
7½	5	3	2	6	2	0	3¾	7½	1¼	4⅜	1¾	4½	2½	6¼
9	5	10	3	0	2	6	4½	9	1⅝	5	2¼	5	3¼	7¾
10½	6	8	3	6	3	0	5¼	10½	1⅞	5¼	2½	5¼	3¾	8¼
12	7	2	4	0	3	0	6	12	2⅛	5½	3	5¾	4½	9
15	7	8	4	6	3	6	7½	12	2½	6½	3½	6¾	5½	9¾
18	10	6	5	0	4	0	9	15	3	7	4	7¼	7	11¼

G. ONE STEAM-CYLINDER AND ONE SINGLE-ACTING RAM WITH SINGLE-ACTING PISTON COMBINED.

MAY AND MOUNTAIN'S IMPROVED RAM AND PISTON-PUMP.

The annexed illustrations represent Messrs. May and Mountain's vertical steam-pump. Figs. 122 and 123 are side and front elevations, and fig. 124 is a sectional elevation.

The steam-cylinder is fitted with a metallic piston of the ordinary kind, and the slide-valve is of the ordinary D type, actuated by an eccentric on the fly-wheel shaft. The standard which carries the cylinder is made hollow, and answers the purpose of an air-vessel. The fly-wheel is mounted midway between the steam-cylinder and the pump, and the crank works in a slot-link connecting the piston-rod and ram, and thereby rendering a connecting-rod unnecessary. It will be observed in the section that the piston-rod is fitted and cottered into a boss on the top side of the slot-link, whilst the ram is cottered to a conical lifting-piece on the bottom of the slot-link. To the bottom of the ram is fitted a piston about twice the area of the ram, and which works in a cylinder, open at the top to the delivery-chamber, A, which is in direct communication with the air-vessel, B. The action of the pump is as follows: Supposing the engine to be at rest

FIG. 122. FIG. 123.

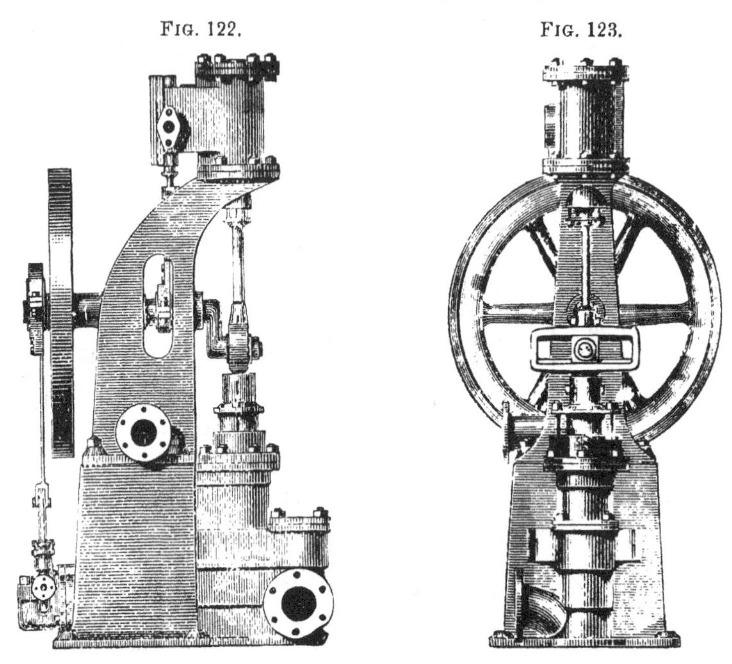

in the position shown, and about to make the up-stroke, as the piston ascends, water enters at the suction-valve, and fills the space below the receding piston. During this time the water filling the space in the cylinder above the piston is discharged into the air-vessel, B, whence it is forced through the discharge-pipe, C. As the piston begins to descend the water, which during the up-stroke entered at the suction-valve, is driven through the delivery-valve into the discharge-chamber, air-vessel, and discharge-pipe.

A feed-pump, as shown in fig. 122, can be supplied at a small extra cost.

FIG. 124.

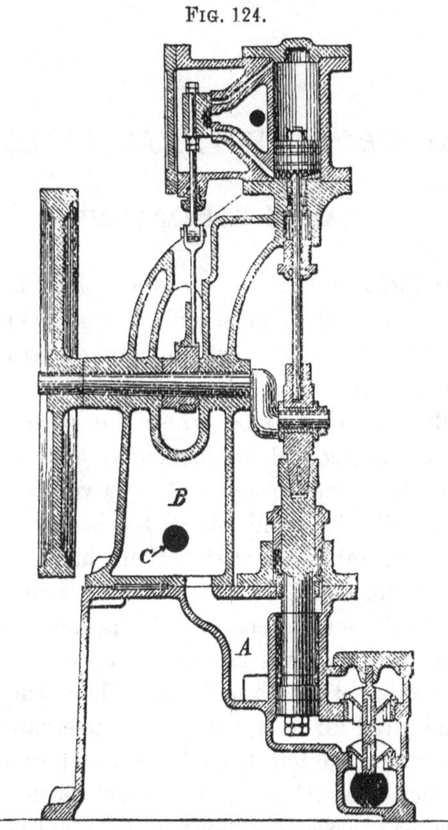

This pump is manufactured in various sizes from a 3-inch steam-cylinder and 2-inch plunger upwards. The suction and delivery-valves are accessible through the same cover or bonnet, the suction-valve being inserted through the delivery valve-seat. The valves are made of the best gun-metal, except when the pump is used for ammoniacal liquor or other fluids that affect the gun-metal, in which case they are made of cast-iron or malleable cast-iron. The piston is fitted with cup leathers when cold water is pumped, but if it is used for hot water, a piston with water-packing grooves is used. For sizes, capacities, and speed see table, page 195.

B. *NON-ROTARY VERTICAL ENGINES.*

SIMPLE STEAM-PUMPS.

THIS class includes such engines as the 'Special,' 'Universal,' and 'Imperial' pumps, embodying the same principles of construction and action as the horizontal pumps of the same names. Their use has been confined to draining in shafts, wells, and other places where the space is very limited and where the horizontal engine could not be conveniently used. They have only been used to a very small extent in mines and collieries, and since he began this work the author has been informed that makers have ceased to push the sale of them, and have discontinued their manufacture. The 'Universal' vertical engine it appears was abandoned about eight or nine years ago in favour of the short-stroke horizontal engine of the same name. The 'Imperial' pump, although admired for its symmetrical appearance, was defective in working, and has been long withdrawn. The withdrawal of the 'Special' pump is of more recent date. It is a matter of some surprise that makers have not succeeded with these pumps. The difficulty with regard to the steam-valve gear, which appears to be the chief source of trouble, could be removed by either balancing the valve by a cataract or a weighted lever or by twisting the steam-ports and placing the valve horizontally,* which would partially remove the objection of the weight of the parts giving an unequal motion for the opening and closing of the ports. Still the fact remains that pumps of this class have been withdrawn from the market.

* This has been done by Messrs. Cope and Maxwell.

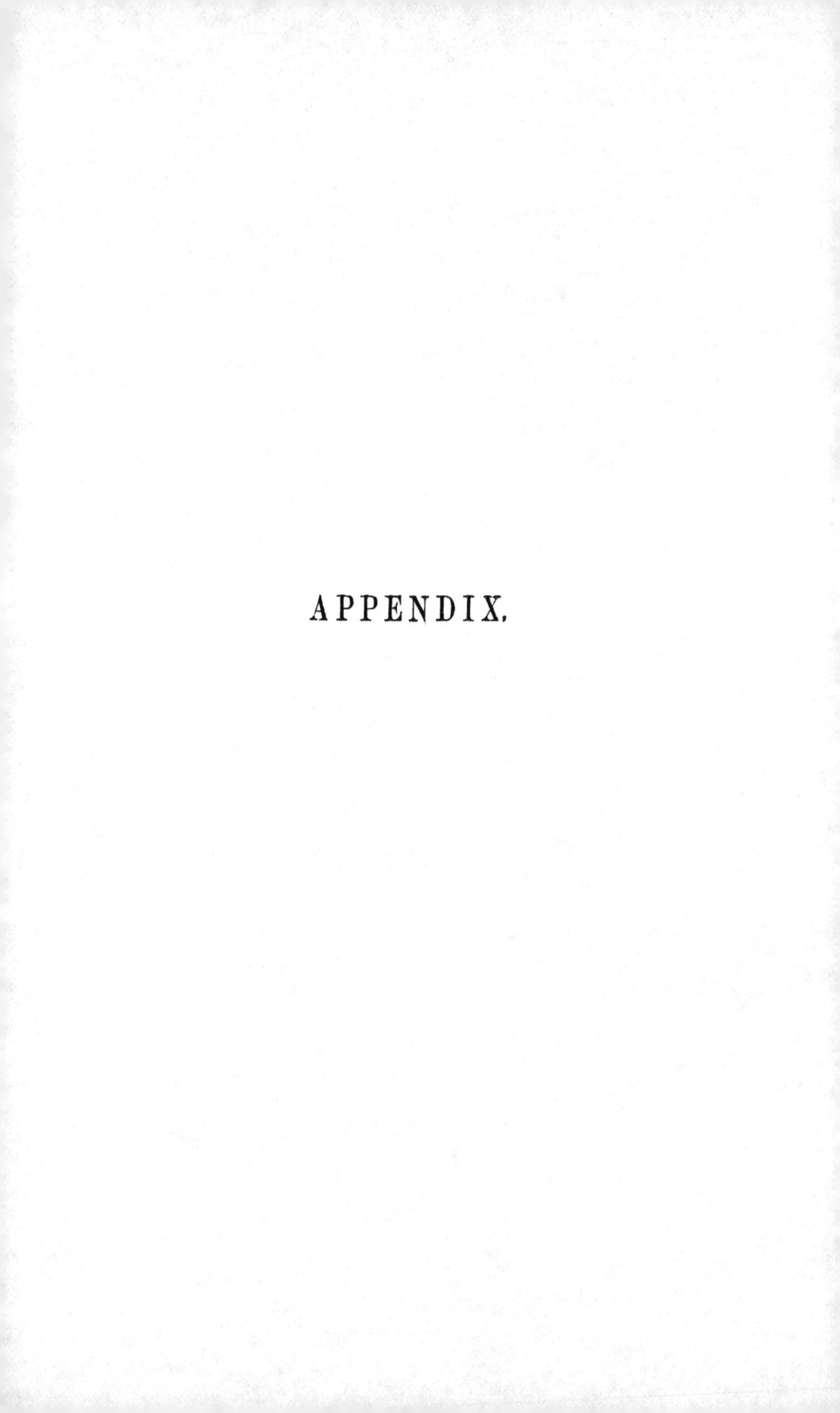

APPENDIX.

HYDRAULIC PUMPING ENGINES.

A NECESSARY and indispensable appendage to an underground pumping engine when used in situations where it would be liable to be submerged if placed within suction reach of the water, by a temporary stoppage of the engine for repairs or by the incursion of a flood into the workings of the mine, is a hydraulic-engine.

The flooding of an underground engine would imperil the safety of a mine unless there be another pumping engine at work, out of reach of the water which has overtaken the underground engine at the bottom of the mine. The necessity of a hydraulic-engine is not so great when the steam-engine is of small or medium size, and therefore practically portable, as it could be removed in a very short time; even if overtaken by the water, it would not be a very difficult task, in many instances, to raise it; but colossal engines—vying in power with Cornish engines—of the type manufactured by Messrs. Hathorn, Davey, and Co., could not be removed. An engine of the latter kind is used in a large chamber specially prepared for it, and the details are so massive that they have to be put together in the chamber. To all intents and purposes such an engine is a stationary one, and it would practically be an impossibility to remove it during submergence; in fact, such a task would not be attempted, unless the abandonment of the mine were intended. Under the circumstances another engine would be set to work, and the water pumped out and the submerged engine recovered in this way. The obtaining, fixing, and setting to work of any other engine would require more or less time according to circumstances. The requisite pumping power might be on the mine or might be obtained from an adjoining mine or an adjacent foundry. In this case the delay would probably be inconsiderable, and the consequences perhaps would not be very serious. But should there be no pumping power thus available, and it has to be obtained from a great distance, a considerable delay must take place, and several days might elapse before an engine could be got to operate upon the water. The loss entailed upon the miner under such circumstances would be a very serious one. Not only is there the stoppage of the output from perhaps the most productive parts of the mine, but a great expense of procuring an additional engine and the liability of damage to the workings underground, and in some instances the loss in this way has been immense. In the case of an inundation in a mine where only a small engine is used, not only would the procuring of another engine be attended with less difficulty, as the smaller engines are very numerous and could either be obtained in a few hours from another mine or from the nearest maker of steam-pumps, but the submerged engine itself might be disconnected under water and drawn up bodily, and be again set to work at a higher level.

For small and medium engines of the simple class the hydraulic engine may generally be dispensed with, as their use necessitates considerable additional power, entails additional cost and trouble, and makes the pumping apparatus of a more complex character. Indeed, it is not a commendable feature of this system of mine drainage that a hydraulic-engine should ever become a necessary part of it. For draining dip workings, in which a steam-pump cannot be conveniently fixed, or where a bad roof renders the engine and steam-pipes dangerous, the hydraulic-engine is indispensable. The principle upon which it works is the raising of a larger quantity of water a less height by the power derived from a smaller quantity of water from a greater height. The power may be obtained from a natural head or from the main pumping engine.

'There are many situations in which a small quantity of water supplied under a considerable pressure can be profitably employed to pump a greater quantity of water against a less pressure. In dip workings in collieries, for instance, engines thus worked can be advantageously used to raise water to the main pumping engines, the motive water in this case being supplied from the rising-main of the main engine. In hilly mining districts, too, water drawn from a

FIG. 125.

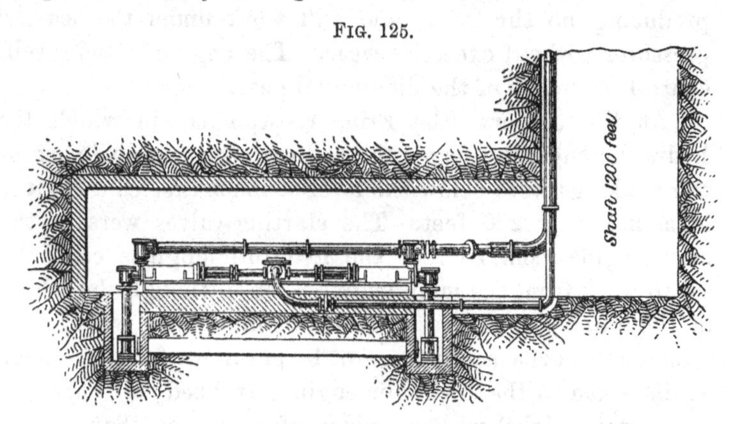

high level may be conducted by pipes into the mines, and then used to raise a greater quantity of water to the surface, thus avoiding the necessity for steam-power.' *

At page 134 in the remarks on 'Compound Steam Pumping Engines,' an account is given of the pumping operations in a colliery 1,200 feet deep, and an explanation of the use of

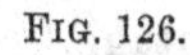
FIG. 126.

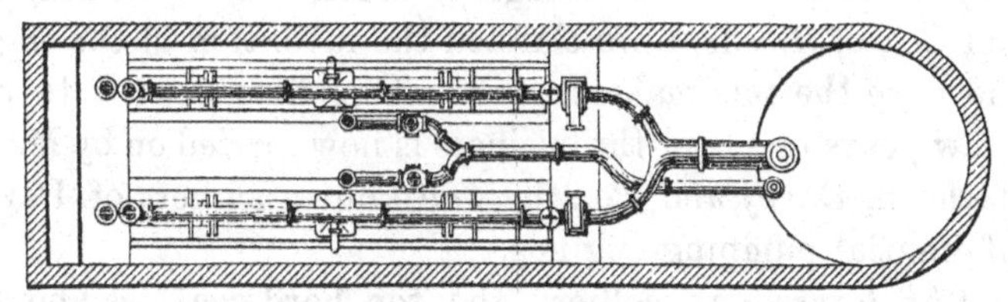

the hydraulic-engine represented by an illustration which clearly shows the mode of application.

The accompanying illustrations, figs. 125 and 126, are an

* 'The Differential and Improved Hydraulic Pumping Engines,' by Henry Davey, M.I.C.E. *Trans. Midland Institute Mining, Civil, and Mechanical Engineers,* March, 1877.

elevation and plan of a pair of Davey's differential hydraulic-engines just referred to, and made to raise 1,000 gallons per minute 300 feet high. In this engine there is no piston, neither is there any internal packing or slide-valve to get out of order. The packing is all applied in stuffing-boxes. The valves are gun-metal mitre-valves, with large areas, producing no throttling, and will work under the heaviest pressures without excessive wear. The engine is under self-control by means of the differential gear.

At the colliery (the Erin, Westphalia) in which the hydraulic-engines referred to were used, the principal parts of the workings were below the level of the main steam-engines by as much as 250 feet. The starting-valves were in the main engine-room, so that the hydraulic-engines could be set to work from the main engine room, although the whole of the lower part of the pit might be filled with water. As a further security a staple might be put down from the main engine-room to the hydraulic-engines (tubbed or otherwise made water-tight) with a ladder of access, so that in the event of a flood in the workings the hydraulic-engines would still be in the dry and accessible for repairs.*

The following remarks on Joy's patent hydraulic pumping engine, with an account of its employment at the Grassmore and Clay Cross collieries, will be interesting as showing the circumstances under which engines of this kind may be used with advantage.† Messrs. Carrett, Marshall, and Co., of Leeds, have claimed the invention of the engine, and were the sole makers of it. This firm ceased to exist a few years ago, and the business is now carried on by Messrs. Hathorn, Davey and Co., the well-known makers of Davey's differential pumping-engines.

'At Grassmore colliery the top hard coal lies about 60 yards below the surface of the ground. The water of this

* 'Direct-Acting Pumping Engines and Pumps,' by Henry Davey, M.I.M.E. *Proceedings of Inst. Mechanical Engineers*, Oct. 29, 1874.

† 'Joy's Patent Hydraulic Pumping Engine,' by Edmund Bromley. *Trans. Chesterfield and Derbyshire Inst. Mining, Civil, and Mechanical Engineers*, July and Oct., 1871.

mine is raised from the sump to "day" by two 15-inch lifts or pumps, of 60 yards each, standing side by side, and worked by a Cornish pumping engine having a 60-inch cylinder with a 10-feet stroke, and 9-feet stroke in the pumps. The dip water was formerly raised by horizontal plunger-pumps placed in the dip workings, which were worked by wire ropes attached to the main-rod of the large pumping engine, and thus forcing the water from the dip workings, up the rising main, into the sump at the bottom of the shaft. Owing to the constant stretching of the ropes, and breakages from that cause, the hindrances became very frequent, and it was found to be necessary to change the system. Accordingly, a pair of hydraulic-engines have been erected, and are now pumping the water. The water lifted by the 15-inch pumps from the sump is delivered into the condensing cistern of the large Cornish engine, for the purpose of supplying the condenser with cold water, and then flows freely away. A portion of this is now taken down again into the mine, from whence it came, to supply the motive power to the hydraulic pumping engine. One end of each pressure-pipe conveying the water down to the hydraulic-engines is projected a few inches into the condensing cistern, about two feet from the top; a large cullender or grate is placed over the end of each to keep out dirt, wood, and ashes, which are sometimes brought up with the pit water. They are taken down the vertical shaft about 60 yards, then turned in the direction of the dip of the coal, and after crossing the levels, one is taken down the engine plane, and the other down the pump-head to the receiver, and from thence to the engines.

'About 100 yards below the engines in the pump-head is the lodge where the dip water is collected, and from whence it is pumped. The suction-pipes are 6 inches diameter, laid in the pump-head. The engines and pumps are placed horizontally, side by side, on the same frame, at a distance of about 660 yards from the sump at the bottom of the shaft. Each pump draws its water from the same suction-pipe, and forces it into the ascending delivery-pipe; each engine exhausts its water into the one delivery-pipe; in other words, the water which propels the engines, and that which is pumped from the lodge, are both together forced up the same rising-main, and delivered into the sump.

'The cylinders and the pumps are of equal capacity, being 8 inches diameter, 18 inches stroke. The pumps are double-acting. The pistons of each cylinder and pump are attached to a common rod. The suction-valves, or clacks, are placed below, in connection with the suction-pipe; and the holding valves immediately over the pumps, and communicating with the rising-main. The cylinder has ports similar to those in a steam-engine; a slide-valve works over the top of the ports; the slide-valve is moved by its connection with a double piston, moving in two short-bored chambers, in the ends of the valve-box. The double pistons are actuated by water-pressure, which is let into and out of the bored chambers alternately by a four-way cock. Motion is given to the four-way cock by a lever and rod, which are coupled to an arm on the piston-rod. This rod is fitted with tappets for adjusting the action of the four-way cock. All the working parts of these engines and pumps are capped with brass. The piston of both cylinders and pumps are similar in construction. Each piston is packed with two leather cups, having a brass collar between them. The whole is bolted down to a strong wood frame. A sluice-valve is placed on the inlet-pipe to turn the water on and off. The slide-valve is made of hard wood, guaiacum or lignum vitæ.

'The action of the engine is as follows: When the sluice-valve is opened on the inlet-pipe, the water enters through the port into the cylinder, and pressing upon the piston, pushes it to the other end. Just before it arrives at that point, the lever of the four-way cock is caught by the tappet which is moved by the piston-rod, the motion is reversed, and the piston is brought back again and so on, alternately. The performance of one of these engines may be expressed thus:—The cylinder 8 inches diameter and 18 inches stroke contains in round numbers 6½ imperial gallons. If 100 gallons of water be taken down from the surface per minute, it will produce 16 strokes per minute of the engine and pump, and force up 100 gallons from the lodge in addition to that taken down to work the engine, making 200 gallons per minute delivered into the sump, the speed of the piston being about 50 feet per minute; or, the duty may be expressed thus: The total vertical head is 258 feet, the length of the pipe through which it passes 2,160 feet. This, making an allowance for friction, gives 108 lbs. per square inch. The height to which the water is delivered is 75 feet, the length of pipe through which it is forced 1,866 feet; this, with the friction added, will give a back pressure of 43 lbs. per square-inch. Then $108-43=65$ lbs. per square-inch available pressure to work the pump. The area of the cylinder$=50{\cdot}26$ inches $\times 65=3267$ lbs. pressure in the cylinder, and $50{\cdot}26 \times 43=2161$ lbs. pressure in the pump, thus giving about 66 per cent. of duty.

'This would give the large pumping engine an extra quantity of 100 gallons per minute, but as its average speed is about three strokes per minute with ample power and pump room, the extra quantity is pumped by increasing the speed less than one extra stroke per minute.

'These engines have been at work night and day alternately for more than two years. They do not appear to be adapted to a high speed. The wear and tear consist mainly in the renewal of the leather packings of the cylinder and pump-pistons. Of course these wear much faster when the water is muddy than when it is clear. They require but little attention, one of the deputies going twice or three times each shift to see that all is right, and to regulate the speed of the engine according to the quantity of water, which varies very much. They are, no doubt, economical engines; they require no fuel nor stoker. There are no moving parts to give friction and to wear out, such as wire rope, wood rods, pulley guides,—except the engine itself. The water power enters silently at one end of the system, and is discharged with little noise at the delivery-pipes; they need but little oil and tallow, as the water itself is the principal lubricator.

'The writer has permission to state that a hydraulic engine of this kind is made, and is shortly to be erected at the Clay Cross works, in one of their underground workings, for the purpose of raising water from the dip workings in the coal to a higher level, whence it is to be conveyed to the standage or sump, and thence lifted to "day" by the large pumping engine. Cylinder 5½ inches diameter, 3-feet stroke, pump 8 inches diameter, 3-feet stroke. The pistons of the cylinder and pump are attached to a common rod. The construction and action of the engine and pump are similar in all respects to the one already described. Messrs. Carrett, Marshall, and Co. are the makers. About 50 feet down the pumping shaft is the piper coal, which gives out a feeder of 45 gallons per minute. This is to be conveyed by a pipe down the vertical shaft 136 feet, and then turned down the engine plane to the hydraulic-engine. The total length is 2,642 feet, with a vertical head of 346 feet. The falling gradient of the plane is about 1 in 12. The water has to be pumped from the suction-well to a water level, a distance of 1,452 feet, and a vertical height of 53 feet. The supply pipe is 4 inches diameter; the suction and delivery pipes 7½ inches diameter.

'The duty of the engine under this arrangement may be stated thus. The total vertical height of the column which actuates the engine is 346 feet, which will give a pressure of 152 lbs. to the square-inch. The vertical height to which the water is to be raised is 53 feet, representing a pressure of 24 lbs. per square-inch. Then $152 - 24 = 128$ lbs. available pressure per square-inch to work the pumps. Then area of 5½ inches $= 23{\cdot}75 \times 128$ lbs. $= 3040$ lbs. pressure in cylinder. Area of 8-inch pump $= 50{\cdot}26 \times 24 = 1206$ lbs. pressure in pump, which gives or shows about 40 per cent. of duty. Or the duty may be expressed thus: The cylinder 5½ inches diameter will contain in round numbers 3 gallons. A feeder of 45 gallons per minute will fill it fifteen times, and the engine will make 7½ strokes per minute. The 8-inch pump will contain 6½ gallons, which being filled 15 times will pump 96 gallons per minute out of the suction-well, which added to

45 gallons used by the engine=141 gallons sent up the delivery-pipe per minute.'

The following account of Mr. Henry Davey's hydraulic-engine, to which we have already referred, is taken from a paper read before the Midland Institute of Engineers by Mr. Davey, Oct. 13, 1875. 'The engine under notice has been designed to obviate the difficulties of wear and tear experienced with hydraulic-engines having pistons and slide-valves, and gearing for working the valves. Pistons and slide-valves wear very rapidly with dirty water, but plungers are practically unaffected. It will be seen that in the engine illustrated there are no pistons, but the power is applied and the work done entirely with plungers. The power-plungers are stationary, and are made to serve as pipes to convey the water from the valve-box (to which they are fixed) to the inside of the pump-plungers; these latter forming the power-cylinders, and being connected to each other by side rods passing outside the valve-box. In this way the forcing stroke of one pump-plunger causes the suction stroke of the other, and *vice versâ*.

'The most novel part of the engine is the valve-box. For hydraulic power no valves answer better than the single mitre-valve; for if these valves get leaky because of sand and grit, they are easily replaced by duplicates, or ground tight again. The difficulty with single mitre-valves of the ordinary type is, however, that of working them, when made of sufficient size to produce but little throttling. This difficulty is entirely obviated in the engine under notice; for the valves, instead of being actuated by means of metallic connections, are worked under water-pressure by means of a small subsidiary-valve acted on by tappets from the engine at the ends of the stroke. On this construction the valves may be made of any size, and a full and free water way given, so as to realise the greatest possible useful effect. By regulating the passages between the subsidiary and main-valves, the latter are made to rise and fall at any required speed, so that the beat may be entirely taken off, and the wear and tear of the valves reduced to a minimum.

Each plunger communicates with a chamber in the valve-box, fitted with one inlet and one exhaust-valve. The valves (which are of gun-metal working on gun-metal seats) have plungers formed on them above and below, the lower plungers being packed with double-cupped leathers as shown, while in the case of the inlet-valves the upper plungers are also packed with cupped leathers. The lower plungers are thus really the rams of small hydraulic presses, and by admitting water under pressure below them the valves are lifted. Diagonal passages connect the plunger-chamber of the right-hand exhaust-valve with the plunger-chamber of the left-hand inlet-valve and *vice versâ*.'

The following is an account of the pumping machinery at the Erin colliery, Westphalia, referred to a few pages back. 'The pit is 1,200 feet deep, and a pair of steam-engines and hydraulic-engines combined are together capable of delivering the required quantity of 1,000 gallons of water per minute, or 60,000 gallons per hour, to the surface. At a point 900 feet below the surface are placed a pair of compound differential pumping engines, with separate condensers. Each engine is capable of raising 500 gallons per minute to the surface, and at the same time of supplying power to a pair of hydraulic pumping engines placed at the bottom of the pit, and employed in raising 500 gallons each per minute to the main engine.

'Each engine has a massive girder bed of cast-iron, with a 35-inch high-pressure and a 60-inch low-pressure cylinder, 6-feet stroke. The slide-valves for these cylinders are actuated by the author's patent differential valve-gear. The pistons are fitted with metallic-packing. The pumps are double-acting ram-pumps, 12½ inches diameter, 6-feet stroke, with rams of gun-metal. The pump-rods, 5¾ inches diameter, pass from end to end of the rams, and are secured thereto by gun-metal nuts. The piston-rods and pump-rods are united in one strong wrought-iron crosshead, supported and guided in cast-iron slide-bars. The pump-valves are of gun-metal, double-beat, 11 inches in diameter, and low lift, allowing ample area in the water-passages, and thus reducing friction to a minimum. The suction-pipes are 10 inches diameter and the rising-main 14 inches diameter; all the joints of the pumps and pipes are of the well-known hydraulic type, made with gutta-percha.

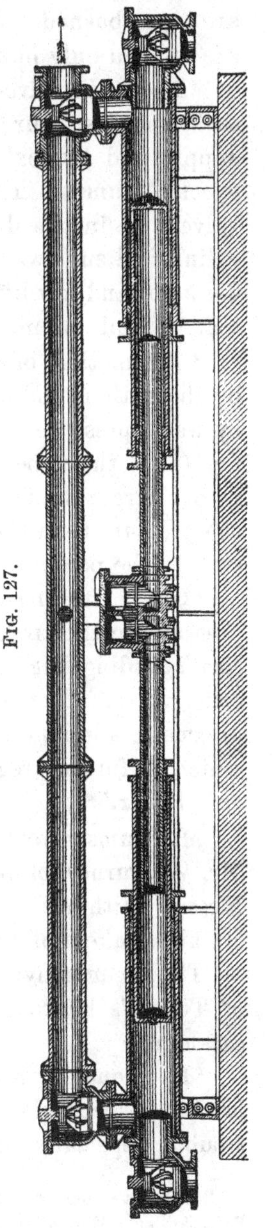

FIG. 127.

'At the bottom of the pit, 300 feet below the main engines, are placed a pair of hydraulic-engines, figs. (125, 126, and 127); these engines lift 1,000 gallons per minute to the main engine, at a speed of twelve strokes per minute, whilst the main engines finally force the water to the surface, and at the same time, through the column, supply power to work the hydraulic-engines. The latter consist of double-acting power-rams and pumps, placed face to face on the telescopic principle. These engines are self-contained on a strong girder-bed similar to that of the steam-engine. The pumps, which are placed at each end of the bed, are 14 inches in diameter, and of the ordinary ram type, with gun-metal double-beat valves, and brass-covered rams 12 inches diameter and 5-feet stroke. The two rams are connected to each other by two strong wrought-iron side rods, and thus alternately pull one another in their return stroke. The power-rams, which are 7 inches diameter, are wholly of gun-metal, and of the same stroke as the pumps. They are connected at one end to a central valve-box, and being cast hollow, admit the pressure to the cylindrical rams of the pumps. All the bearings and glands are brass-bushed. All the packings are easy of access, and are applied only in stuffing-boxes.

'The main valve-box is placed in the centre of the engine, and contains four valves, 5¾ inches diameter, two being supply and two exhaust, which communicate with their respective pumps. The valves are cylindrical gun-metal mitre valves, 6½ inches diameter at top, and provided at their cylindrical ends with double cup leathers. They are of large diameter and low lift, thus reducing friction and wear and tear to a minimum.

'The method of actuating these valves has been patented by the author, and consists in lifting and closing the valve by water-pressure. Thus in the arrangement illustrated in fig. (127) the upper or cylindrical portion of the valve is of rather larger diameter than the valve proper, consequently the pressure from the main lifts this valve and admits pressure to the power-ram.

'On the return stroke this valve is closed by admitting pressure on the large area at the top of the valve, whilst a corresponding but intermittent action takes place with the relief-valve. The main-valves thus rise and fall by the alternate admission and cutting off the water-pressure, which is further regulated by a modification of the differential gear.' *

Mr. Ramsbottom, hydraulic engineer, of Hunslet, Leeds, Mr. W. Turner, of Adelphi Street, Salford, Manchester, and Messrs. Withinshaw and Co., of Wiggin Street, Birmingham, are also makers of hydraulic-engines.

The accompanying illustration (fig. 128) and description of Turner's hydraulic-engine are taken from Mr. Turner's catalogue:

'It frequently happens in collieries that small quantities of water accumulate down brow, which give considerable trouble, especially in such cases where a bad roof renders

* From a paper on 'The Underground Pumping Machinery at the Erin Colliery, Westphalia,' read before the Society of Engineers, by Mr Henry Davey, on June 12, 1876.

Fig. 128

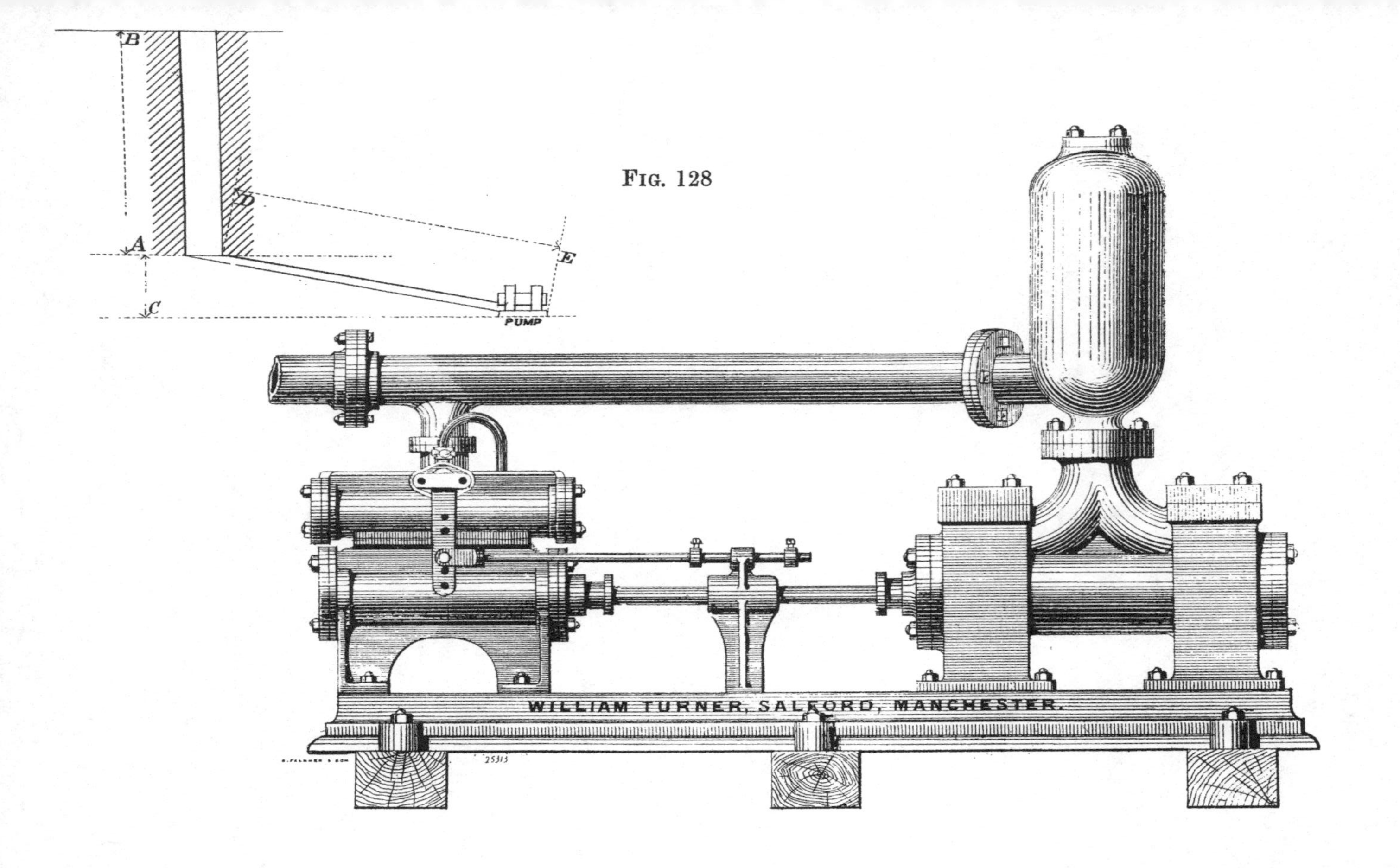

the use of the steam-engine and steam-piping objectionable, and often dangerous. The hydraulic-engine has been designed to overcome this difficulty.

'The principle upon which it works is as follows, viz.: The pressure obtained from a small quantity of water from a high level is able to raise a greater quantity of water to a less head, and the size of the power cylinder is so proportioned that the engine will not only deliver to the required head the quantity of water due to the lift-pump, but also the quantity of water which has been sent down to work the engine. The power may be obtained from a natural head, or from the main pumping engines, suitable pressure and delivery piping being carried down to the engine.

'The working parts are entirely of gun-metal, the pistons being fitted with double cup leathers.

'Should the pressure be great, or the water very dirty, the power-cylinder and pumps are arranged with double rams.

'These pumps are so simple in construction and require so little attention, that they will continue working although completely under water.

'When inquiring for particulars it will be necessary to state the quantity of water to be lifted per hour, the natural head or depth of pit (A B), the dip (A C), and the distance (D E).'

PUMP-VALVES.

When it is considered that with unsuitable or bad valves in the pump no engine can be an efficient one, and can give no satisfactory working result for the force exerted, it is not surprising that for a great many years the most careful attention has been given to the subject of pump-valves, and many varieties have been devised differing not only in form but in the materials of which they are constructed. In this instance necessity has been the prolific mother of a very numerous family, many of which have passed away after a very doubtful existence of a few months, others have lingered out a few years in obscurity, and have gradually become lost to sight, whilst others again, after attracting a little notice in the world of mining, and receiving a small share of patronage, have been eventually pushed aside by the later and superior offspring of the same mythical parent, so that through one cause or another of a host of valves which have from time to time made their appearance only a comparatively small number have stood the severe test of work in a mine or colliery, and many of these have acquired a local rather than a general reputation, and are only to be found in certain mining districts.

Of course in such a number of valves differing in many cases so widely from each other, there are qualities which are peculiar to the different valves, adapting them for use in certain kinds of engines and under certain circumstances. No valve therefore can be the most useful under every circumstance.

A copious description of pump-valves and a minute consideration of their construction and merits, such as are contained in Messrs. Michell and Letcher's Essay on 'Cornish Mine Drainage,' * and such as will be found in Mr. P. R. Björ-

* See 'Royal Cornwall Polytechnic Society's Report for 1875.'

ling's large work on the subject now in course of preparation, and which the author anticipates with much interest, will not be permissible in these pages. All he purposes to do is to give a short account of some of the best known varieties of pump-valves, and their merits with respect to underground steam-pumping engines. The construction of many of the valves that will be noticed, and the principle involved therein, are familiar to most mining engineers. We shall therefore dispense with illustrations, and give only an outline description of some of the valves. Pump-valves may be reduced to two groups, namely, *Metallic Valves* and *India-rubber Valves*. Without entering into minutiæ or regarding subtle distinctions or taking account of a few valves which are more curious than useful, the first group may be divided into two and the second group into three classes, as shown below.

I. Metallic Valves.

1. *Clack-Valves.*

(*a*) Ordinary Clack-Valves.
(*b*) Clack-Valves, with contrivances for facilitating exit of water.

2. *Valves with Vertical Lift.*

(*a*) Single-Beat Valves.
(*b*) Double-Beat Valves.
(*c*) Treble-Beat Valves.
(*d*) Quadruple-Beat Valves.
(*e*) Six-Beat Valves.

II. India-Rubber Valves.

(1) *Flat Valves.*
(2) *Conical Valves.*
(3) *Lip-Valves.*

I. METALLIC VALVES.

1. Clack-Valves.

(a) *Ordinary Clack-Valves.*

These are valves which are hinged to the seat either with rivets or bolts or with a pin, each end of which works in a recess or slot. In the latter case the valve lifts a little at the hinge. There are several varieties of these valves known in the Cornish mines, where scarcely any other kinds are used, as the Butterfly, Hake's Mouth, Jan Ham's, Trelease's. They are used very extensively; in the Cornish pump-lifts, and wherever the Cornish system of pumping prevails, they are almost exclusively used. They are sluggish in their action, and the concussion in closing is very great, owing to the high lift, and they are not suitable for steam-pumps, as they allow too much slip of water, which in short-stroke engines, particularly at high speeds, would lead to a very serious loss of useful effect, amounting in some instances to 10 per cent. of the water pumped.

(b) *Clack-Valves, with contrivances for facilitating exit of Water.*

Teague's noiseless valve consists of the ordinary Hake's Mouth variety, with the central portion removed and a small valve hung over the opening. Mr. Teague is also the originator of a clack-valve with a series of perforations over which circular pieces of india-rubber work (on a spill working through the central hole of each series). Jenkyn's valve consists of three clack-valves, over each other. Valves on the principle of Mr. Teague's noiseless valve have given much satisfaction, and they are a great improvement upon the ordinary clack-valves, and ought to supersede them in heavy pump-lifts. These valves are not suitable for pumps running a large number of strokes per minute, as they do not close with sufficient rapidity. Valves with the lowest lift, such as those on the principle of the Cornish double-beat and quadruple-beat valves (Harvey and West's and Husband's) require springs, under such circumstances, to expedite their closing.

2. Valves with a Vertical Lift.

(a) *Single-Beat Valves.*

This class comprises cup-valves, wing- or mitre-valves, spill- or mushroom-valves, and ball-valves. The cup-valve is a short cylinder closed at the top, which slides within the seat, and has vertical orifices. The wing-valve consists of a flat or rounded circular piece, surmounting a number of radial arms or wings which act as guides. In the spill-valve the valve is carried by a spill which works through the centre of the seat. The ball-valve is a metallic ball which works within a framework to prevent its being thrown out of its place. All these valves have a high lift. With the exception of the ball-valve they are suitable for light lifts. For working at a high speed they should be fitted with an india-rubber spring, and have a large area in proportion to the working-barrel to reduce the lift. For mine or colliery pumps up to 10 inches diameter the best valves are the wing- or mitre-valves, made of gun-metal with lignum vitæ beats, and an india-rubber spring on the top. In his 'Salford' pump, Mr. Turner uses a mushroom-valve. Mitre-valves, with rubber springs, are used in the Self-Governing engines of Messrs. Hayward Tyler and Co.; mitre-valves are also used in some varieties of the 'Universal' pump, and in the 'Caledonia,' and Messrs. Parker and Weston's, and Turner's engines of the same class.

Messrs. Tangye Brothers and Holman use in the 'Special' pump a brass valve of the mushroom type; the guide (which passes through the valve) is also of brass. They are fitted with a spring consisting of a tube of india-rubber, the bottom end of which encloses the top of the valve. The section of the 'Special' pump shows the arrangement. In

Colebrook's pump wing-valves, of the form shown in one of the illustrations accompanying the account of that pump, are used.

(*b*) *Double-Beat Valves.*

The best variety of these valves are Harvey and West's, too well known to need description. These valves have a large lifting area, spacious waterway, and consequently a low lift; there is therefore little concussion, and the slip of water is very little. They are largely used in waterworks, but are not suitable for ordinary sinking pump-lifts, owing to the solid impurities in the water. They are not used in Cornwall. Where the water is clean and free from grit they are the best valves. For a high rate of speed they should be fitted with an india-rubber spring to secure a prompt closing. Mr. Henry Davey states that valves (Harvey and West's) with the bottom beats of gun-metal and the top ones of hippopotamus hide, have been at work for twelve months under heads of 600 to 1,000 feet in dirty water without new leather beats being required. Valves of this kind are used in Davey's 'Differential' engines. Mr. Davey has found by experiment that the slip with these valves is 7½ per cent. less than with the ordinary butterfly-valves. Rhinoceros hide is said to be very durable, and when let into the seats of the ordinary double-beat valve to make one of the best and most durable of valves, but it is expensive. Fig. 136 is a section of this valve, with a rubber spring, and shows one way of securing the seat.

(*c*) *Treble-Beat Valves.*

Valves on the same principle as that of Harvey and West's have been constructed with three beats. Messrs. Carnell and Hosking introduced a valve of this class many years ago, but it is not now in use, as far as the author is aware. They are not suitable for mines or collieries.

(*d*) *Quadruple-Beat Valves.*

Husband's four-beat valve is merely a further extension of the principle carried out in the last two classes named.

It is well adapted to waterworks purposes, but is not suitable for mine or colliery work. 'Some of these engines' (direct-acting) 'used for pumping water for the supply of towns make the up-stroke of 10 feet in a second with poles reaching a diameter of 50 inches, and this valve allows the water to pass with such freedom that the valve is closed at the moment the up-stroke is completed, and there is consequently no loss of water or shock. About sixteen years since the writer made valves on this principle for a pump for pumping water from a depth of 320 yards in one lift, at the Eagle Bank Colliery, Yorkshire, and they acted without shock under that great head of water, and are at work to the present day.' *

(e) *Six-Beat Valves.*

Mr. J. Simpson has designed a valve formed of three rings or annular valves, the inner and outer edges of which have beats. Each valve is separate, and the lift is independent of the others. The valves are arranged in conical order (the bottom being the largest and the top smallest).

II. INDIA-RUBBER VALVES.

1. *Flat Valves.*

These valves consist of a flat circular piece of india-rubber fastened at the centre to a grating or grid, which forms the seat of the valve. They are useful for pressures not exceeding fifty pounds.

Mr. George Massey is the inventor of an excellent valve of this class with a revolving disc, which has been used successfully for heads of 350 feet (about 150 lbs. per square-inch). The inventor thinks they would stand greater pressures. The rotative motion, which is caused by oblique waterways, equalises the wear and promotes the durability of the valve.

* 'Pumping Machinery,' by Mr. W. Husband, C.E., *Proceedings Mining Inst. of Cornwall*, vol. i. No. 5, p. 167.

Mr. Massey has also designed a valve with four small revolving discs on one grid. To accelerate the closing india-rubber springs are sometimes applied to the flap. Mr. Holman has designed a valve consisting of a rectangular grid covered with round bars of india-rubber fastened at the ends with glands. Mr. Thomson is the designer of a novel flat-valve, in which the grid is dispensed with. There is an annular piece of india-rubber, which first gives exit to the water by the bending upwards of its edges, and as the pressure increases the valve is lifted bodily.

The Imperial Steam Pump Company use Massey's disc-valves in their pumps, and Messrs. Hayward Tyler and Co. formerly used disc-valves.

2. *Conical Valves.*

These valves consist of a perforated cone wound around which are bands or rings of india-rubber. In some valves the water escapes by distension of the india rubber, in others by its compression. Mr. Morris's pyramid or hatband-valve consists of several short cylinders, in one casting, diminishing gradually from the bottom to the top. Around each cylinder is a band of india-rubber, fastened at the bottom by an iron clamp. They can be used for pressures not exceeding 200 feet. The collective areas of the holes are one and a-half times the area of the pump. The respective pumps of Mr. Holman, Mr. West, and Mr. Morrison, are on the same principle, but differ in construction. In Hosking's ball-valve a series of india-rubber balls are arranged in circles on a cone. Some valves are a combination of the disc and the cone—the former surmounting the latter.

3. *Lip-Valves.*

These are made wholly of india-rubber. The principle involved is the forcing open of a slit in a cone or the forcing apart of two or more pieces of rubber where their edges meet by the pressure of the water and the closing of the same by the natural elasticity of the material.

Field's disc lip-valve is composed of two flat or slightly coned india-rubber discs placed face to face horizontally. The central portion is removed, and the inner edge of each piece is firmly fixed into a groove in the main body of the pump, in such a way as to cause the discs to impinge at their outer edges. The arrangement may be fairly conceived by placing two saucers together, the top one being inverted. The water enters through the orifice in the centre of the bottom disc, and finds an exit by forcing open the discs at the outer edges. Mr. Field has also designed a compound-valve on the same principle, consisting of three lips. Perreaux's valve is a flanged tube of india-rubber flattened at the top, thus forming lips.

India-rubber valves generally. India-rubber valves are not suitable for high lifts. The best varieties are not useful for lifts above 350 feet, and most of them can only be safely used up to 100 feet. For pumps with a lively action and working under a low pressure, india-rubber valves have been used to advantage. Messrs. Hayward Tyler and Co. employ india-rubber ball-valves in some of their 'Universal' pumps. They wear very evenly, and are useful for pressures not exceeding 100 lbs. per square-inch. Field's and Perreaux's valves may be used for low pressures, and for dirty or thick fluids, but the wear of the latter is very great, and the cost of maintenance is large.

JOINTS OF MAIN PIPES.

It is very important that the engineer and miner should pay great attention to the joints of the pipes of long lifts, as without a suitable joint there is likely under great pressures to be a leakage. In Cornish mines, where the lift

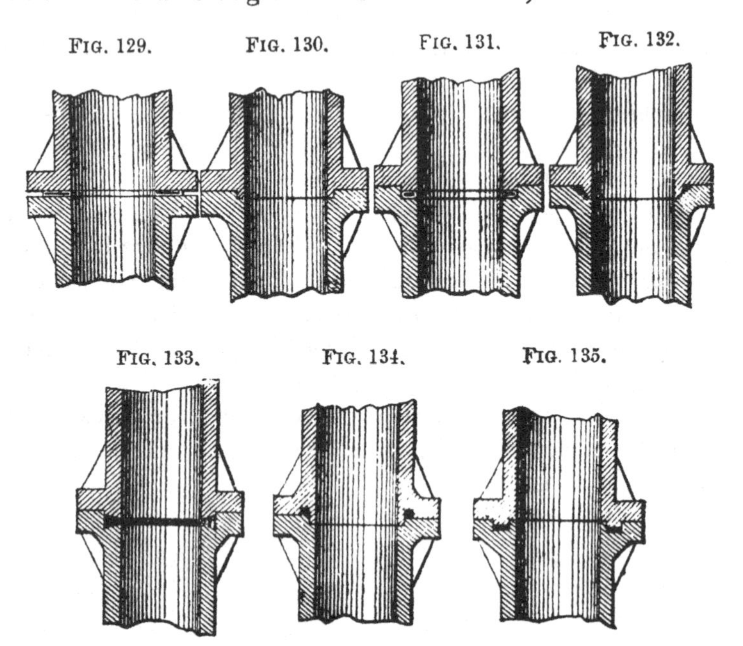

rarely exceeds 300 feet (equal to a pressure of about 130 lbs. per square-inch at the bottom), the joint shown at fig. 129 is quite sufficient. The unplaned faces of the flanges are brought together, with a thin iron ring covered with tarred flannel inserted betwen them, and bolted in the ordinary way. If the flanges be planed they may be made to set so exactly that no rings are needed. A more elaborate joint, fig. 130, is sometimes used in Cornish lifts in collieries. A similar joint is shown at fig. 131, which is used for the pipes of the 'Universal' steam-pump. A lead ring, wrapped around with lamp cotton, without tar, is bolted between the two flanges. They have been tested to 500 lbs. per square-inch, or say 1,000 feet head of water. Another joint used occasionally for the same pump is that of fig. 132. One pipe is recessed, or 'bell-mouthed,' as one says in the North, and the other has a projecting ring to fit the bell-mouth, which is chamfered at the top, as shown by the black part. A round cord of gutta-percha is cut off to the required length to go round the pipe, indicated by the black, and the ends are melted together by a piece of hot iron, thus forming a ring. When the flange bolts are tightened, the gutta-percha is squeezed into the shape shown in the sketch.

Sir William Armstrong has, with great advantage, made his joints, which have to stand a pressure of 700 lbs. per square-inch as shown in fig. 133. A cord of gutta-percha, shown in black, is inserted between the pipe flanges, the lower of which is recessed, and the upper has a corresponding rim or projection. The pressure of the bolts squeezes the gutta-percha flat. It will be noticed that the rim is slightly bevelled, the inner edge projecting beyond the outer. Sir William Armstrong has laid down miles of pipes with this kind of joint, without testing them.

At Nixon's Navigation Colliery, South Wales, this kind of joint is used in connection with the 'Differential' pumping-engine. It is used almost exclusively for heavy pressures by Messrs. Hathorn, Davey, and Co., of Leeds. The gutta-percha ring is also used for making the joints of valve-box covers; the ring adheres to the cover, and the latter may be taken off and replaced almost any number of times without requiring a new ring. In the joint shown in fig. 134, it will be noticed that the flange of the upper pipe terminates in a nipple or projection, with a groove turned around it. An india-rubber ring is put into the groove. The flange of the lower pipe has a recess turned out to correspond with the projection in the upper flange. In Fig. 135 the india-rubber is placed in a groove in the lower flange.

WIPPERMAN AND LEWIS'S PATENT AUTOMATIC PUMP, FOR SUPPLYING AIR VESSELS WITH AIR.

A VERY useful apparatus for all pumping machinery fitted with air-vessels is that illustrated at figs. 136 and 137, and which we will introduce in the words of the maker, Mr. William Turner, of Salford:

'The necessity of a regular supply of air to the air-vessels of all pumping engines has led to the introduction of various systems of force-pumps and separate air-chambers, which have all been more or less failures, either from an inadequate supply of air or being too cumbersome in application. The accompanying design illustrates Wipperman and Lewis's apparatus for this purpose, as applied to an ordinary pumping engine; it consists of a cylindrical vessel (A), fig. (136), which has no working parts, the water itself forming the piston: at the bottom of the chamber is a small pipe (B), fitted with a regulating cock (C), which is attached to the pump valve-box, immediately below the delivery-valve; at the top of the vessel is fixed a small gun-metal valve-box, fitted with inlet and outlet air-valves, and from this a delivery-pipe communicates directly to the air-vessel.

'The action of the apparatus is as follows, viz.: When the main-pump draws its water it will partly empty the vessel A, the amount being indicated by the gauge F, and regulated to a nicety by the cock C; on the return stroke of the pump-plunger, the whole of the air drawn into the chamber A is sure to be delivered into the air-vessel G, because the pressure in the main-pump, when delivering, is in all cases greater than on the suction side.

'The advantages of this simple contrivance over others are many: The supply can be regulated as in no other

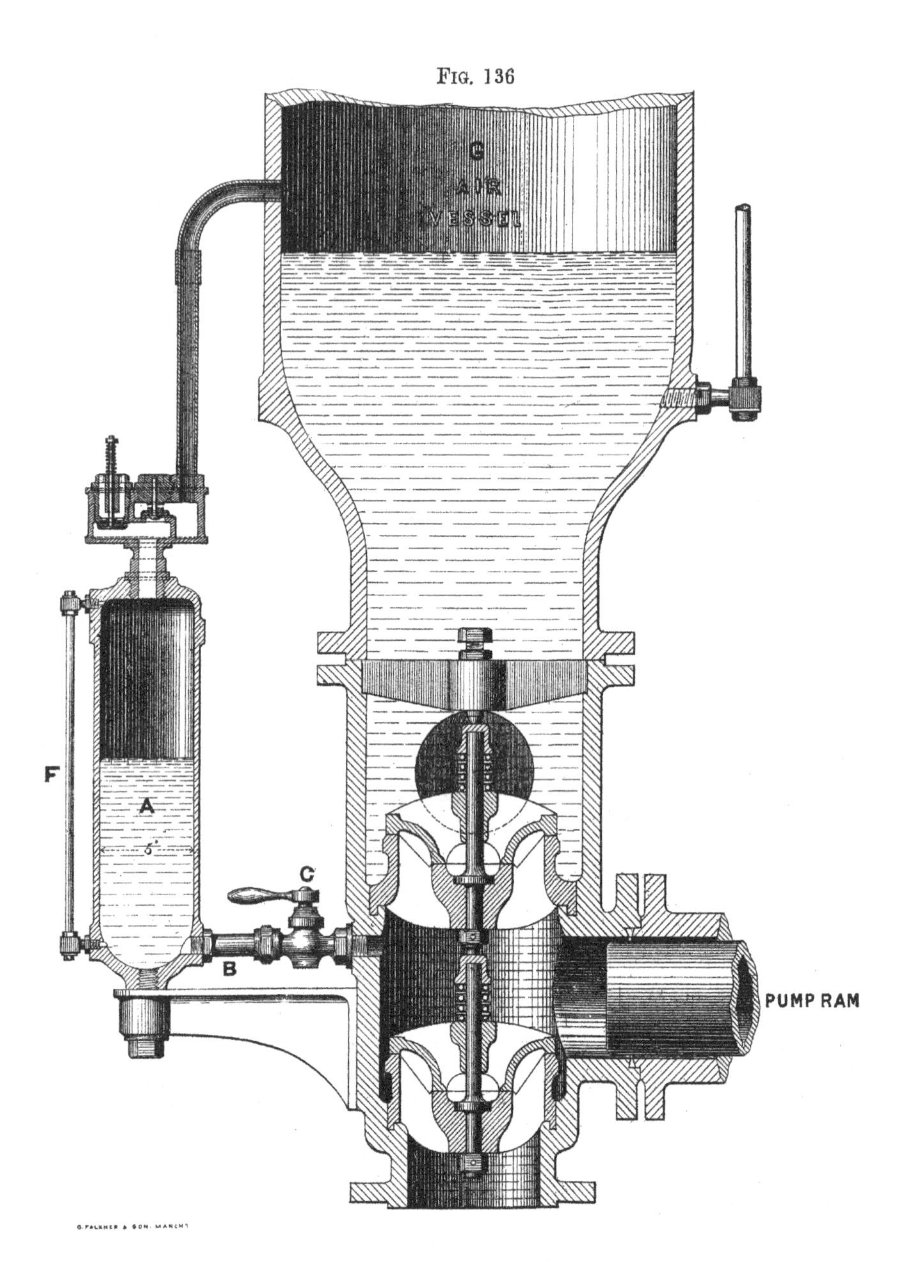

Fig. 136

Fig. 137

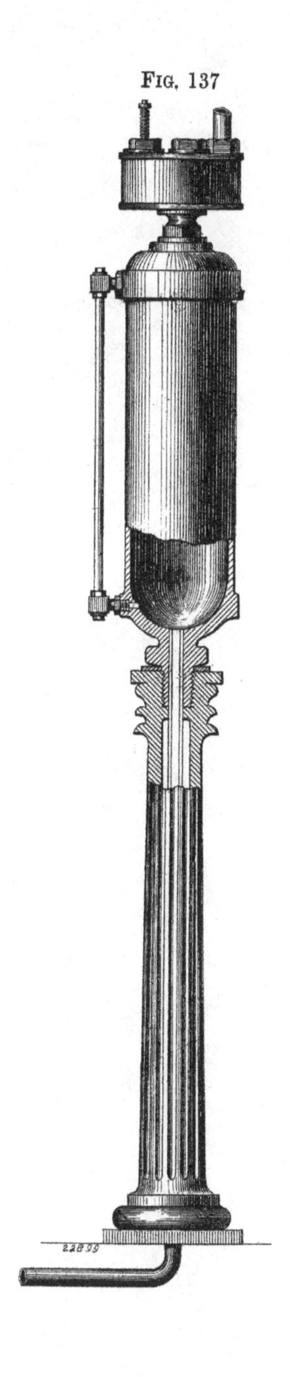

apparatus; there can be no friction, as there are no working parts; it is perfectly noiseless in action, and may be attached to any engine, whatever the construction, at one-tenth the cost of an ordinary charging-pump; it may also form an ornament to the engine-room (see arrangement on column, fig. 137), and may be placed in any convenient position entirely away from the engine, providing its connection to the pump and air-vessel be arranged as indicated. The apparatus may also be applied to all machines where a constant and regular supply of air is required.

'On inquiry for particulars it will be necessary to state the quantity of water, the pressure or vertical head, and number of strokes per minute.'

	£	s.	d.
A Size	10	0	0
B ,, ...	12	0	0
C ,, ...	14	0	0
D ,, ...	16	0	0

HYDRAULIC MEMORANDA.

Weight of Water at its Common Temperature.

1	cubic	inch	=·03617 lb. (·5778 oz.); (253·182 grains).
12	,,	inches	=·434 lb.
1	,,	foot	= 62·5 lbs. (1000 ozs.); (6·25 gallons).
1·8	,,	foot	= 1 cwt. (11·2 gallons).
35·84	,,	feet	= 1 ton (224 gallons).
276·5	,,	inches	= 1 gallon.
1	cylindrical	inch	=·02842 lb.
12	,,	inches	=·341 lb.
1	,,	foot	= 49·1 lbs (4·91 gallons).
2·282	,,	feet	= 1 cwt.
45·64	,,	,,	= 1 ton.
1	lb. avoirdupois		=27·712 cubic inches.

I. *To find the pressure per square inch of a column of water.**

Multiply the height in feet by ·434.

The pressure per circular inch may be found by multiplying the height in feet by ·341.

Example.—Required the pressure in pounds per square inch of a column of water 200 feet high.

$$200 \times \cdot434 = 86\cdot8 \text{ lbs. per square inch.}$$

A ready way of ascertaining approximately the pressure is to take half the height in feet. The difference is on the side of safety.

IA. *To find the pressure of a column of water in pounds.**

If the base be circular, square the diameter in inches and multiply by ·341 or ·34, which gives the weight of one foot in height; therefore by multiplying by the number of feet in height the pressure is found.

* From Mr. Telford's 'Memorandum Book.'

If the base be square, multiply by ·434.

Examples.—Required the pressure of a column of water 12 inches in diameter and 20 feet high.

$12 \times 12 \times ·341 \times 20 = 982·080$ lbs. if the base be circular.

$12 \times 12 \times ·434 \times 20 = 1249·920$ lbs. if the base be square.

II. *To find the quantity of water in a pipe.**

The square of the diameter in inches gives the weight of water in pounds for 3 feet in length, and by striking off one figure to the right the number of gallons is found. Rule III. is an accurate method for determining quantities.

Example.—Required the quantity of water which a pipe 15 inches in diameter and 9 feet long will contain.

$15 \times 15 \times 3 = 675$ lbs., or 67·5 gallons.

For large cylinders, where both the dimensions are in feet, the following rule of the Rev. W. N. Griffin † is easily calculated and is sufficiently accurate for ordinary use. Multiply the square of the diameter by five times the depth, the result being the number of gallons approximately. By deducting 2 per cent., the exact quantity may be more nearly ascertained. The actual contents of a cylinder in gallons is found by multiplying the square of the diameter in feet by the depth in feet and then by 4·8974, which is nearly 5.

Example.—Required the quantity of water which a cylinder 4 feet diameter and 5 feet long will contain.

$4 \times 4 \times 25 = 400$ gallons, or after 2 per cent. has been deducted = 392. The actual contents are 391·79 gallons.

If either or both dimensions be in inches the result has to be divided a proper number of times by 12. A pipe 3 inches diameter and 3 feet long holds $\frac{15}{16}$ of a gallon or $7\frac{1}{2}$ pints roughly; after taking off 2 per cent., 7·35 pints. The actual contents are 7·346 pints.

$3 \times 3 \times 15 = 135 \div 144 = ·938$ gallons.

* The area of the internal diameter of a cylinder multiplied by its depth, equals its cubical capacity.

† See *Engineer* of Aug. 9, 1878.

III. *To find the quantity of water in gallons delivered by a pump per minute.*

Square the diameter in inches and multiply by ·034 and the product by the speed in feet.*

Example.—Required the quantity of water delivered by a pump 10 inches in diameter and working at the rate of 40 feet per minute.

$$10 \times 10 \times \cdot 034 \times 40 = 136 \text{ gallons per minute.}$$

TABLE I.—*Showing pressure of water in pounds per square inch for various heights.*

Height in feet	Pressure per sq. inch in lbs.	Height in feet	Pressure per sq. inch in lbs.	Height in feet	Pressure per sq. inch in lbs.
1	·434	35	15·19	270	117·18
2	·868	40	17·36	280	121·52
3	1·302	45	19·53	290	125·86
4	1·736	50	21·70	300	130·20
5	2·170	55	23·87	310	134·54
6	2·604	60	26·04	320	138·88
7	3·038	65	28·21	330	143·22
8	3·472	70	30·38	340	147·56
9	3·906	75	32·55	350	151·90
10	4·340	80	34·72	360	156·24
11	4·774	85	36·89	370	160·58
12	5·208	90	39·06	380	164·92
13	5·642	95	41·23	390	169·26
14	6·076	100	43·40	400	173·60
15	6·510	110	47·74	450	195·30
16	6·944	120	52·08	500	217·00
17	7·378	130	56·42	550	238·70
18	7·812	140	60·76	600	260·40
19	8·246	150	65·10	650	282·10
20	8·680	160	69·44	700	303·80
21	9·114	170	73·78	750	325·50
22	9·548	180	78·12	800	347·20
23	9·982	190	82·46	850	368·90
24	10·416	200	86·80	900	390·60
25	10·850	210	91·14	950	412·30
26	11·284	220	95·48	1000	434·00
27	11·718	230	99·82		
28	12·152	240	104·16		
29	12·586	250	108·50		
30	13·020	260	112·84		

* The capacity of a cylinder 1 inch in diameter, and 1 foot in length, equals 034 of a gallon.

TABLE II.—*Showing quantity of water in gallons delivered at each stroke by various sizes of pumps, from 3 to 27 inches in diameter, and strokes from 1 inch to 9 feet.*

PUMP.	LENGTH OF THE STROKE.										
Diameter in Inches.	1″	5″	6	7″	8″	9″	10″	11″	12″	13″	14″	15″
3	·0255645	·127	·153	·178	·204	·230	·255	·281	·306	·332	·357	·383
4	·0454481	·227	·272	·318	·363	·409	·454	·499	·545	·590	·636	·681
5	·0710126	·355	·426	·497	·568	·639	·710	·781	·852	·923	·994	1·065
6	·1022582	·511	·613	·715	·818	·920	1·022	1·124	1·227	1·329	1·431	1·533
7	·1391848	·695	·835	·974	1·113	1·252	1·391	1·531	1·670	1·809	1·948	2·087
8	·1817924	·908	1·090	1·272	1·454	1·636	1·817	1·999	2·181	2·363	2·545	2·726
9	·2300810	1·150	1·380	1·610	1·840	2·070	2·300	2·530	2·760	2·991	3·221	3·451
10	·2840506	1·420	1·704	1·988	2·272	2·556	2·840	3·124	3·408	3·692	3·976	4·260
11	·3437012	1·718	2·062	2·405	2·749	3·093	3·437	3·780	4·124	4·468	4·811	5·155
12	·4090329	2·045	2·454	2·863	3·272	3·681	4·090	4·499	4·908	5·317	5·726	6·135
13	·4800455	2·400	2·880	3·360	3·840	4·320	4·800	5·280	5·760	6·240	6·720	7·200
14	·5567392	2·783	3·340	3·897	4·453	5·010	5·567	6·124	6·680	7·237	7·794	8·351
15	·6391139	3·195	3·834	4·473	5·112	5·752	6·391	7·030	7·669	8·308	8·947	9·586
16	·7271696	3·635	4·363	5·090	5·817	6·544	7·271	7·998	8·726	9·453	10·180	10·907
17	·8209063	4·104	4·925	5·746	6·567	7·388	8·209	9·029	9·850	10·671	11·492	12·313
18	·9203240	4·601	5·521	6·442	7·362	8·282	9·203	10·123	11·043	11·964	12·884	13·804
19	1·0254227	5·127	6·152	7·177	8·203	9·228	10·254	11·279	12·305	13·330	14·355	15·381
20	1·1362025	5·681	6·817	7·953	9·089	10·225	11·362	12·498	13·634	14·770	15·906	17·043
21	1·2526632	6·263	7·515	8·768	10·021	11·273	12·526	13·779	15 031	16·284	17·537	18·789
22	1·3748050	6·874	8·248	9 623	10·998	12·373	13·748	15·122	16·497	17·872	19·247	20·622
23	1·5026278	7·513	9·015	10·518	12 021	13·523	15 026	16·528	18·031	19·534	21·036	22·539
24	1·6361316	8·180	9·816	11·452	13·089	14·725	16·361	17·997	19·633	21·269	22·905	24·541
25	1·7753164	8·876	10 651	12 427	14·202	15·977	17·753	19·528	21·303	23·079	24·854	26 629
26	1·9201822	9·600	11 521	13·441	15·361	17·281	19·201	21·121	23·042	24·962	26·882	28·802
27	2·0707291	10·353	12·424	14·495	16·565	18·636	20·707	22·778	24·848	26·919	28·990	31·060

TABLE II.—*continued.*

PUMP.	LENGTH OF THE STROKE.												
Diameter in Inches.	16″	17″	18″	21″	2′ 0″	2′ 3″	2′ 6″	2′ 9″	3′ 0″	3′ 3″	3′ 6″	3′ 9″	4′ 0″
3	·409	·434	·460	·536	·613	·690	·766	·843	·920	·997	1·073	1·150	1·227
4	·727	·772	·818	·954	1·090	1·227	1·363	1·499	1·636	1·772	1·908	2·045	2·181
5	1·136	1·207	1·278	1·491	1·704	1·917	2·130	2·343	2·556	2·769	2·982	3·195	3·408
6	1·636	1·738	1·840	2·147	2·454	2·760	3·067	3·374	3·681	3·988	4·294	4·601	4·908
7	2·226	2·366	2·505	2·922	3·340	3·758	4·175	4·593	5·010	5·428	5·845	6·263	6·680
8	2·908	3·090	3·272	3·817	4·363	4·908	5·453	5·999	6·544	7·089	7·635	8·180	8·726
9	3·681	3·911	4·141	4·831	5·521	6·212	6·902	7·592	8·282	8·973	9·663	10·353	11·043
10	4·544	4·828	5·112	5·965	6·817	7·669	8·521	9·373	10·225	11·077	11·930	12·782	13·634
11	5·499	5·842	6·186	7·217	8·248	9·279	10·311	11·342	12·373	13·404	14·435	15·466	16·497
12	6·544	6·953	7·362	8·589	9·816	11·043	12·270	13·498	14·725	15·952	17·179	18·406	19·633
13	7·680	8·160	8·640	10·080	11·521	12·961	14·401	15·841	17·281	18·721	20·161	21·602	23·042
14	8·907	9·464	10·021	11·691	13·361	15·031	16·702	18·372	20·042	21·712	23·383	25·053	26·723
15	10·225	10·864	11·504	13·421	15·338	17·256	19·173	21·090	23·008	24·925	26·842	28·760	30·677
16	11·634	12·361	13·089	15·270	17·452	19·633	21·815	23·996	26·178	28·359	30·541	32·722	34·904
17	13·134	13·955	14·776	17·239	19·701	22·164	24·627	27·089	29·552	32·015	34·478	36·940	39·403
18	14·725	15·645	16·565	19·326	22·087	24·848	27·609	30·370	33·131	35·892	38·653	41·414	44·175
19	16·406	17·432	18·457	21·533	24·610	27·686	30·762	33·838	36·915	39·991	43·067	46·144	49·220
20	18·179	19·315	20·451	23·860	27·268	30·677	34·086	37·497	40·903	44·311	47·720	51·129	54·537
21	20·042	21·295	22·547	26·305	30·063	33·821	37·579	41·337	45·095	48·853	52·611	56·369	60·127
22	21·996	23·371	24·746	28·870	32·995	37·119	41·244	45·368	49·492	53·617	57·741	61·866	65·990
23	24·042	25·544	27·047	31·555	36·063	40·570	45·078	49·586	54·094	58·602	63·110	67·618	72·126
24	26·178	27·814	29·450	34·358	39·267	44·175	49·083	53·992	58·900	63·809	68·717	73·625	78·534
25	28·405	30·180	31·955	37·281	42·607	47·933	53·259	58·585	63·911	69·237	74·563	79·889	85·215
26	30·722	32·643	34·563	40·323	46·084	51·844	57·605	63·366	69·126	74·887	80·647	86·408	92·168
27	33·131	35·202	37·273	43·485	49·697	55·909	62·121	68·334	74·546	80·758	86·970	93·182	99·394

TABLE II.—*continued.*

PUMP.	LENGTH OF THE STROKE.											
Diameter in Inches.	4′ 3″	4′ 6″	4′ 9″	5′ 0″	5′ 6″	6′ 0″	6′ 6″	7′ 0″	7′ 6″	8′ 0″	8′ 6″	9′ 0″
3	1·303	1·380	1·457	1·533	1·687	1·840	1·994	2·147	2 300	2·454	2·607	2·760
4	2 317	2·454	2 590	2·726	2·999	3·272	3·544	3·817	4·090	4·363	4·635	4·908
5	3 621	3·834	4·047	4·260	4·686	5 112	5 538	5·965	6 391	6·817	7·243	7·669
6	5 215	5·521	5·828	6 135	6·749	7·362	7·976	8·589	9·203	9·816	10·430	11·043
7	7·098	7·515	7·933	8·351	9·186	10 021	10·856	11·691	12·526	13·361	14·196	15 031
8	9·271	9·816	10·362	10·907	11·998	13·089	14·179	15·270	16·361	17·452	18·542	19·633
9	11·734	12·424	13·114	13·804	15·185	16 565	17·946	19·326	20·707	22·087	23·468	24·848
10	14·486	15·338	16·190	17·043	18·747	20·451	22·155	23·860	25·564	27·268	28·973	30·677
11	17·528	18·559	19·590	20·622	22·684	24·746	26·808	28·870	30·933	32·995	35·057	37·119
12	20·860	22·087	23·314	24·541	26·996	29·450	31·904	34·358	36·812	39·267	41·721	44·175
13	24·482	25·922	27·362	28·802	31·683	34·563	37·443	40·323	43·204	46·084	48·964	51·844
14	28·393	30·063	31·734	33·404	36·744	40·085	43·425	46·766	50·106	53·446	56·787	60·127
15	32·594	34·512	36·429	38·346	42·181	46·016	49·850	53·685	57·520	61·354	65·189	69·024
16	37·085	39·267	41·448	43·630	47·993	52·356	56·719	61·082	65·445	69·808	74·171	78·534
17	41·866	44·328	46·791	49·254	54·179	59·105	64·030	68·956	73·881	78·807	83·732	88·657
18	46·936	49·697	52·458	55·219	60·741	66·263	71·785	77·307	82·829	88·351	93·873	99·394
19	52·296	55·372	58·449	61·525	67·677	73·830	79·982	86·135	92·288	98·440	104·593	110·745
20	57·946	61·354	64·763	68·172	74·989	81·806	88·623	95·441	102·258	109·075	115·892	122·709
21	63·885	67·643	71·401	75·159	82·675	90·191	97·707	105·223	112·739	120·255	127·771	135·287
22	70·115	74·239	78·363	82·488	90·737	98·985	107·234	115·483	123·732	131·981	140·230	148·478
23	76·634	81·141	85·649	90·157	99·173	108·189	117·204	126·220	135·236	144·252	153·268	162·283
24	83·442	88·351	93·259	98·167	107·984	117·801	127·618	137·435	147·251	157·068	166·885	176·702
25	90·541	95·867	101·193	106·518	117·170	127·822	138·474	149·126	159·778	170·430	181·082	191·734
26	97·929	103·689	109·450	115·210	126·732	138·253	149·774	161·295	172·816	184·337	195·858	207·379
27	105·607	111·819	118·031	124·243	136·668	149·092	161·516	173·941	186·365	198·789	211·214	223·638

The gallon in this Table is reckoned as containing 276·5 cubic inches. No allowance has been made for slip of water through the valves, which varies according to the nature of the valves and the action of the engine. A column is included showing, to several decimals, the number of gallons for one inch of stroke in order that the delivery for any length stroke not given may be calculated from it and with the greatest accuracy, as may also that of the lengths of stroke given. It contains very useful data for anyone who has to make calculations of this kind, and any degree of exactness can be obtained, according to the number of decimals used.

TABLE III.—*Showing quantity of water in gallons delivered per minute by various sizes of pumps, from 3 to 27 inches in diameter, at speeds varying from 1 foot to 300 feet.*

Diameter of Pump in Inches.	PISTON SPEED IN FEET PER MINUTE.									
	1	30	40	50	60	70	80	90	100	110
3	·306774	9·203	12·270	15·338	18·406	21·474	24·541	27·609	30·677	33·745
4	·545377	16·361	21·815	27·268	32·722	38·176	43·630	49·083	54·537	59·991
5	·852151	25·564	34·086	42·607	51·129	59·650	68·172	76·693	85·215	93·736
6	1·227098	36·812	49·083	61·354	73·625	85·896	98·167	110·438	122·709	134·980
7	1·670217	50·103	66·808	83·510	100·213	116·915	133·617	150·319	167·021	183·723
8	2·181503	65·445	87·260	109·075	130·890	152·705	174·520	196·335	218·150	239·965
9	2·760972	82·829	110·438	138·048	165·658	193·268	220·877	248·487	276·097	303·706
10	3·408607	102·258	136·344	170·430	204·516	238·602	272·688	306·774	340·860	374·946
11	4·124415	123·732	164·976	206·220	247·464	288·709	329·953	371·197	412·441	453·685
12	4·908394	147·251	196·335	245·419	294·503	343·587	392·671	441·755	490·839	539·923
13	5·760546	172·816	230·421	288·027	345·632	403·238	460·843	518·449	576·054	633·660
14	6·630870	200·426	267·234	334·043	400·852	467·660	534·469	601·278	668·087	734·895
15	7·669367	230·031	306·774	383·468	460·162	536·855	613·549	690·243	766·936	843·630
16	8·726035	251·781	349·041	436·301	523·562	610·822	698·082	785·343	872·603	959·863
17	9·850375	295·526	394·035	492·543	591·052	689·561	788·070	886·578	985·087	1083·596
18	11·043888	331·316	441·755	552·194	662·633	773·072	883·511	993·949	1104·388	1214·827
19	12·305073	369·152	492·202	615·253	738·304	861·355	984·405	1107·456	1230·507	1353·558
20	13·634430	409·032	545·377	681·721	818·065	954·410	1090·754	1227·098	1363·443	1499·787
21	15·031959	450·958	601·278	751·597	901·917	1052·237	1202·556	1352·876	1503·195	1653·515
22	16·497660	494·929	659·906	824·883	989·859	1154·836	1319·812	1484·789	1649·766	1814·742
23	18·031534	540·946	721·261	901·576	1081·892	1262·207	1442·522	1622·838	1803·153	1983·468
24	19·633579	589·007	785·343	981·678	1178·014	1374·350	1570·686	1767·022	1963·357	2159·693
25	21·303797	639·113	852·151	1065·189	1278·227	1491·265	1704·303	1917·341	2130·379	2343·417
26	23·042187	691·265	921·687	1152·109	1382·531	1612·953	1843·374	2073·796	2304·218	2534·640
27	24·848749	745·462	993·950	1242·437	1490·925	1739·412	1987·900	2236·387	2484·875	2733·362

TABLE III.—*continued.*

Diameter of Pump in Inches.	PISTON SPEED IN FEET PER MINUTE.									
	120	130	140	150	160	170	180	190	200	210
3	36·812	39·880	42·948	46·016	49·083	52·151	55·219	58·287	61·354	64·422
4	65·445	70·899	76·352	81·806	87·260	92·714	98·167	103·621	109·075	114·529
5	102·258	110·779	119·301	127·822	136·344	144·865	153·387	161·908	170·430	178·951
6	147·251	159·522	171·793	184·064	196·335	208·606	220·877	233·148	245·419	257·690
7	200·426	217·128	233·830	250·532	267·234	283·936	300·639	317·341	334·043	350·745
8	261·780	283·596	305·411	327·226	349·041	370·856	392·671	414·486	436·301	458·116
9	331·316	358·926	386·536	414·145	441·755	469·365	496·974	524·584	552·194	579·804
10	409·032	443·118	477·204	511·291	545·377	579·463	613·549	647·635	681·721	715·807
11	494·929	536·173	577·418	618·662	659·906	701·150	742·394	783·638	824·883	866·127
12	589·007	638·091	687·175	736·259	785·343	834·426	883·510	932·594	981·678	1030·762
13	691·265	748·870	806·476	864·081	921·687	979·292	1036·898	1094·503	1152·109	1209·714
14	801·704	868·513	935·321	1002·130	1068·939	1135·747	1202·556	1269·365	1336·174	1402·982
15	920·324	997·017	1073·711	1150·405	1227·098	1303·792	1380·486	1457·179	1533·873	1610·567
16	1047·124	1134·384	1221·644	1308·905	1396·165	1483·425	1570·686	1657·946	1745·207	1832·467
17	1182·105	1280·613	1379·122	1477·631	1576·140	1674·648	1773·157	1871·666	1970·175	2068·683
18	1325·266	1435·705	1546·144	1656·583	1767·022	1877·460	1987·899	2098·338	2208·777	2319·216
19	1476·608	1599·659	1722·710	1845·760	1968·811	2091·862	2214·913	2337·963	2461·014	2584·065
20	1636·131	1772·475	1908·820	2045·164	2181·508	2317·853	2454·197	2590·541	2726·886	2863·230
21	1803·835	1954·154	2104·474	2254·793	2405·113	2555·433	2705·752	2856·072	3006·391	3156·711
22	1979·719	2144·695	2309·672	2474·649	2639·625	2804·602	2969·578	3134·555	3299·532	3464·508
23	2163·784	2344·099	2524·414	2704·730	2885·045	3065·360	3245·676	3425·991	3606·306	3786·622
24	2356·029	2552·365	2748·701	2945·036	3141·372	3337·708	3534·044	3730·380	3926·715	4123·051
25	2556·455	2769·493	2982·531	3195·569	3408·607	3621·645	3834·683	4047·721	4260·759	4473·797
26	2765·062	2995·484	3225·906	3456·328	3686·749	3917·171	4147·593	4378·015	4608·437	4838·859
27	2981·850	3230·337	3478·825	3727·312	3975·800	4224·287	4472·775	4721·262	4969·750	5218·237

TABLE III.—*continued.*

Diameter of Pump in Inches.	PISTON SPEED IN FEET PER MINUTE.								
	220	230	240	250	260	270	280	290	300
3	67·490	70·558	73·625	76·693	79·761	82·828	85·896	88·964	92·032
4	119·982	125·436	130·890	136·344	141·798	147·251	152·705	158·159	163·613
5	187·473	195·994	204·516	213·037	221·559	230·080	238·602	247·123	255·645
6	269·961	282·232	294·503	306·774	319·045	331·316	343·587	355·858	368·129
7	367·447	384·149	400·852	417·554	434·256	450·958	467·660	484·362	501·065
8	479·931	501·746	523·561	545·377	567·192	589·007	610·822	632·637	654·452
9	607·413	635·023	662·633	690·243	717·852	745·462	773·072	800·681	828·291
10	749·893	783·979	818·065	852·151	886·237	920·323	954·409	988·496	1022·582
11	907·371	948·615	989·859	1031·103	1072·347	1113·592	1154·836	1196·080	1237·324
12	1079·846	1128·930	1178·014	1227·098	1276·182	1325·266	1374·350	1423·434	1472·518
13	1267·320	1324·925	1382·531	1440·136	1497·741	1555·347	1612·952	1670·558	1728·163
14	1469·791	1536·600	1603·408	1670·217	1737·026	1803·834	1870·643	1937·452	2004·261
15	1687·260	1763·954	1840·648	1917·341	1994·035	2070·729	2147·422	2224·116	2300·810
16	1919·727	2006·988	2094·248	2181·508	2268·769	2356·029	2443·289	2530·550	2617·810
17	2167·192	2265·701	2364·210	2462·718	2561·227	2659·736	2758·245	2856·753	2955·262
18	2429·655	2540·094	2650·533	2760·972	2871·410	2981·849	3092·288	3202·727	3313·166
19	2707·116	2830·166	2953·217	3076·268	3199·318	3322·369	3445·420	3568·471	3691·521
20	2999·574	3135·918	3272·263	3408·607	3544·951	3681·296	3817·640	3953·984	4090·329
21	3307·030	3457·350	3607·670	3757·989	3908·309	4058·628	4208·948	4359·268	4509·587
22	3629·485	3794·461	3959·438	4124·415	4289·391	4454·368	4619·344	4784·321	4949·298
23	3966·937	4147·252	4327·568	4507·883	4688·198	4868·514	5048·829	5229·144	5409·460
24	4319·387	4515·723	4712·058	4908·394	5104·730	5301·066	5497·402	5693·737	5890·073
25	4686·835	4899·873	5112·911	5325·949	5538·987	5752·025	5965·063	6178·101	6391·139
26	5069·281	5299·703	5530·124	5760·546	5990·968	6221·390	6451·812	6682·234	6912·656
27	5466·725	5715·212	5963·700	6212·187	6460·675	6709·162	6957·650	7206·137	7454·624

The gallon in this Table is reckoned as containing 276·5 cubic inches. No allowance has been made for slip of water through the valves, which varies according to the nature of the valves and the action of the engine. A column is included showing, to several decimals, the number of gallons for one foot in order that the delivery for any speed not given may be calculated from it, and with the greatest accuracy, as may also that of the speeds given. It contains very useful data for anyone who has to make calculations of this kind, and any degree of exactness can be obtained according to the number of decimals used.

TABLE IV.—*Showing approximately the number of gallons per hour delivered by various sizes of pumps.*

(*Extracted from Messrs. Hayward Tyler and Co.'s Catalogue.*)

Diameter of cylinder	Diameter of pump	Strokes per minute	Strokes per minute	Strokes per minute	Strokes per minute	Strokes per minute	Strokes per minute	Strokes per minute	Strokes per minute
in.	in.	30	40	50	60	70	80	90	100
4	2	225	300	375	450	525	600	675	750
4	3	525	700	875	1,050	1,225	1,400	1,575	1,750
4	4	938	1,250	1,562	1,875	2,188	2,500	2,812	3,125
5	3	700	933	1,166	1,400	1,633	1,866	2,100	2,333
5	4	1,250	1,666	2,083	2,500	2,916	3,333	3,750	4,166
5	5	1,960	2,613	3,266	3,920	4,573	5,227	5,880	6,533
6	3	700	933	1,166	1,400	1,633	1,866	2,100	2,333
6	4	1,250	1,666	2,083	2,500	2,916	3,333	3,750	4,166
6	5	1,960	2,613	3,266	3,910	4,573	5,227	5,880	6,533
6	6	2,810	3,760	4,700	5,640	6,580	7,520	8,460	9,400
7	4	1,250	1,666	2,083	2,500	2,916	3,333	3,750	4,166
7	5	1,960	2,613	3,266	3,920	4,573	5,227	5,880	6,533
7	6	2,820	3,760	4,700	5,640	6,580	7,520	8,460	9,400
7	7	3,840	5,120	6,400	7,680	8,960	10,240	11,520	12,800
7	8	5,010	6,680	8,350	10,020	11,670	13,360	15,030	16,700
9	4	1,575	2,100	2,625	3,150	3,675	4,200	4,725	5,250
9	5	2,450	3,266	4,083	4,900	5,716	6,533	7,350	8,166
9	6	3,525	4,700	5,875	7,050	8,225	9,328	10,494	11,660
9	7	4,800	6,400	8,000	9,600	11,200	12,800	14,400	16,000
9	8	6,240	8,320	10,400	12,280	14,560	16,640	18,720	20,800
9	10	9,750	13,000	16,250	19,600	22,800	26,000	29,000	32,500
9	12	14,000	18,500	23,500	28,000	32,500	37,000	42,300	47,000
12	3	1,050	1,400	1,750	2,100	2,450	2,800	—	—
12	3½	1,440	1,900	2,400	2,880	3,290	3,800	—	—
12	4	1,870	2,500	3,120	3,740	4,300	5,000	—	—
12	4½	2,400	3,200	4,000	4,800	5,600	6,400	—	—
12	5	2,940	3,920	4,900	5,880	6,800	7,800	—	—
12	5½	3,500	5,500	5,900	7,000	8,200	9,400	—	—
12	6	4,200	5,600	7,000	8,400	9,800	11,200	—	—
12	6½	4,950	6,600	8,250	9,900	11,500	13,200	—	—
12	7	5,760	7,680	9,600	11,520	13,440	15,360	—	—
12	7½	6,600	8,800	11,000	13,200	15,400	17,600	19,800	22,000
12	8	7,500	10,000	12,500	15,000	17,500	20,000	22,500	25,000
12	9	9,390	12,520	15,650	18,780	21,910	25,040	28,780	31,300
12	10	11,775	14,800	19,625	23,550	27,475	29,600	35,325	39,250
12	11	14,250	19,000	23,750	28,500	33,250	38,000	42,750	47,500
12	12	16,950	22,600	28,250	33,900	39,550	45,200	50,850	56,500
12	15	26,505	35,340	44,175	53,010	61,845	70,680	79,515	88,350
15	4	2,337	3,128	3,900	4,674	5,450	6,256	—	—
15	4½	3,000	4,000	5,000	6,000	7,000	8,000	—	—
15	5	3,675	4,900	6,120	7,350	8,574	9,800	—	—
15	5½	4,444	5,925	7,400	8,888	10,368	11,850	—	—
15	6	5,250	7,000	8,750	10,500	12,250	14,000	—	—
15	6½	6,187	8,250	10,312	12,374	14,436	16,500	—	—

TABLE IV.—*continued.*

Diameter of cylinder	Diameter of pump	Strokes per minute	Strokes per minute	Strokes per minute	Strokes per minute	Strokes per minute	Strokes per minute	Strokes per minute	Strokes per minute,
in.	in.	30	40	50	60	70	80	90	100
15	7	7,200	9,600	12,000	14,400	16,800	19,200	—	—
15	7½	8,250	11,000	13,750	16,500	19,250	22,000	—	—
15	8	9,375	12,500	15,624	18,750	21,874	25,000	—	—
15	9	11,922	15,896	19,870	23,844	27,818	31,792	—	—
15	10	14,718	19,624	24,530	29,436	34,342	39,248	—	—
15	11	17,820	23,760	29,700	35,640	41,580	47,520	—	—
15	12	21,187	28,250	35,312	42,375	49,437	56,500	—	—
15	15	33,150	44,200	55,250	66,300	77,350	88,400	—	—
18	4	2,790	3,720	4,650	5,580	6,510	7,440	—	—
18	4½	3,600	4,800	6,000	7,200	8,400	9,600	—	—
18	5	3,960	5,280	6,600	7,820	9,240	10,560	—	—
18	5½	5,310	7,080	8,850	10,620	12,390	14,160	—	—
18	6	6,330	8,440	10,550	12,660	14,770	16,880	—	—
18	6½	7,410	9,880	12,350	14,820	17,290	19,760	—	—
18	7	8,640	11,520	14,400	17,280	20,160	23,040	—	—
18	7½	9,900	13,200	16,500	19,800	23,100	26,400	—	—
18	8	11,280	15,040	18,800	22,560	26,320	30,080	—	—
18	9	14,310	18,080	23,850	28,620	33,390	36,160	—	—

TABLE IV.—*continued.*

Diameter of cylinder	Diameter of pump	Strokes per minute	Strokes per minute	Strokes per minute	Strokes per minute	Strokes per minute	Strokes per minute	Strokes per minute	Strokes per minute
in.	in.	25	30	35	40	45	50	55	60
18	10	14,700	17,640	20,580	23,520	26,460	29,400	32,340	35,280
18	11	17,800	21,360	24,920	28,480	32,040	35,600	39,160	42,720
18	12	21,175	25,410	29,645	33,880	38,115	42,350	46,585	50,820
18	15	33,150	39,720	46,410	53,040	59,670	66,300	72,930	79,440
21	4½	3,500	4,200	4,900	5,600	6,300	7,000	7,700	8,400
21	5	4,285	5,142	5,999	6,856	7,713	8,570	9,427	10,284
21	5½	5,165	6,198	7,231	8,264	9,297	10,330	11,363	12,396
21	6	6,125	7,350	8,575	9,800	11,025	12,250	13,375	14,700
21	6½	7,220	8,664	10,108	11,550	12,996	14,440	15,884	17,328
21	7	8,315	9,978	11,641	13,304	14,967	16,630	18,293	19,956
21	7½	9,625	11,550	13,475	15,400	17,325	19,250	21,175	23,100
21	8	10,935	13,122	15,309	17,496	19,683	21,870	24,057	26,244
21	8½	12,425	14,910	17,395	19,880	22,365	24,850	27,335	29,820
21	9	13,780	16,536	19,292	22,048	24,804	27,560	30,316	33,072
21	10	17,060	20,472	23,884	27,296	30,708	34,120	37,532	40,944
21	11	21,000	25,200	29,400	33,600	37,800	42,000	46,200	50,400
21	12	24,720	29,663	34,506	39,549	44,492	49,437	54,380	59,325
21	15	38,675	46,410	54,145	61,880	69,615	77,350	85,085	92,820
21	18	55,650	66,780	77,910	89,040	100,170	111,300	122,430	133,560
26	5	5,290	6,348	7,406	8,464	9,522	10,580	11,638	12,696
26	5½	6,400	7,680	8,860	10,240	11,520	12,800	14,080	15,360
26	6	7,600	9,120	10,640	12 160	13,680	15,200	16,720	18,240
26	6½	8,925	10,710	12,495	14,280	16,065	17,850	19,635	21,420
26	7	10,350	12,420	14,490	16,560	18,630	20,700	22,770	24,840
26	7½	11,875	14,250	16,625	19,000	21,375	23,750	26,125	28,500
26	8	13,550	16,260	18,970	21,680	24,400	27,100	29,810	32,520
26	8½	15,300	18,360	21,420	24,480	27,540	30,600	33,660	36,720
26	9	17,175	20,610	24,045	27,480	30,915	34,350	37,785	41,220
26	10	21,175	25,410	29,645	33,880	38,115	42,340	46,585	50,820
26	11	25,650	30,780	35,910	41,040	46,170	51,300	56,430	61,560
26	12	30,500	36,600	42,700	48,800	54,900	61,000	67,100	73,200
26	15	47,500	57,000	66,500	76,000	85,500	95,000	104,500	114,000
30	6	8,800	10,560	12,320	14,080	15,840	17,600	19,360	21,120
30	6½	10,300	12,360	14,420	16,480	18,540	20,600	22,660	24,720
30	7	12,000	14,400	16,800	19,200	21,600	24,000	26,400	28,800
30	7½	13,750	16,500	19,250	22,000	24,750	27,500	30,250	33,000
30	8	15,675	21,945	18,810	25,080	28,215	31,350	34,485	43,890
30	8½	17,700	21,240	24,780	28,320	31,860	35,400	38,940	42,480
30	9	19,875	23,850	27,825	31,800	35,775	39,750	43,725	47,700
30	10	24,525	29,430	34,335	39,240	44,145	49,050	53,955	58,860
30	11	29,675	35,610	41,545	47,480	53,415	59,350	65,285	71,220
30	12	35,300	42,360	49,420	56,480	63,540	70,600	77,660	84,720
30	15	55,200	66,240	77,280	88,320	99,360	110,400	121,440	132,480

TABLE V.—*Showing approximately the number of gallons delivered per hour, and the height in feet, by various sizes of pumps at different pressures of steam.*

(*Extracted from Messrs. Hayward Tyler and Co.'s Catalogue.*)

Diameter of steam-cyl.	Diameter of pump	Steam 25 lbs.		Steam 30 lbs.		Steam 35 lbs.		Steam 40 lbs.		Steam 45 lbs.		Steam 50 lbs.	
		Gallons per hour	Vertical height in feet	Gallons per hour	Vertical height in feet	Gallons per hour	Vertical height in feet	Gallons per hour	Vertical height in feet	Gallons per hour	Vertical height in feet	Gallons per hour	Vertical height in feet
4	2	750	100	700	120	650	140	600	160	550	180	500	200
4	3	1,750	45	1,650	54	1,550	63	1,450	72	1,350	81	1,250	90
4	4	3,000	25	2,850	30	2,700	35	2,550	40	2,400	45	2,250	50
5	3	2,000	70	1,900	84	1,800	98	1,700	112	1,600	126	1,500	140
5	4	3,500	40	3,340	48	3,180	56	3,020	64	2,860	72	2,700	80
5	5	5,500	25	5,200	30	4,900	35	4,600	40	4,300	45	4,000	50
6	3	2,000	100	1,900	120	1,800	140	1,700	160	1,600	180	1,500	200
6	4	3,500	58	3,340	70	3,180	81	3,020	92	2,860	103	2,700	114
6	5	5,500	35	5,200	42	4,900	49	4,600	56	4,300	63	4,000	70
6	6	7,900	25	7,400	30	6,900	35	6,400	40	5,900	45	5,400	50
7	4	3,500	75	3,310	90	3,180	115	3,020	120	2,860	135	2,700	150
7	5	5,500	50	5,200	60	4,900	70	4,600	80	4,300	90	4,000	100
7	6	7,900	35	7,400	42	6,900	49	6,400	56	5,900	63	5,400	70
7	7	10,800	25	10,200	30	9,600	35	9,000	40	8,400	45	7,800	50
7	8	14,000	19	13,400	23	12,800	26	12,200	30	11,600	34	11,000	38
9	4	3,500	125	3,340	150	3,180	175	3,020	200	2,860	225	2,700	250
9	5	5,500	80	5,200	96	4,900	112	4,600	128	4,300	144	4,000	160
9	6	7,900	55	7,400	66	6,900	77	6,400	88	5,900	99	5,400	110
9	7	10,800	40	10,200	48	9,600	56	9,000	64	8,400	72	7,800	80
9	8	13,700	30	13,200	36	12,700	42	12,200	48	11,700	54	11,200	60
9	10	21,500	20	20,700	25	19,900	30	19,100	35	18,300	40	17,500	45
9	12	31,000	12	29,800	15	28,600	18	27,400	20	26,200	22	25,000	25
12	3	1,750	400	1,650	480	1,550	560	1,450	640	1,350	720	1,250	800
12	3½	2,400	300	2,260	360	2,120	420	1,980	480	1,840	540	1,700	600
12	4	3,000	250	2,850	300	2,700	350	2,550	400	2,400	450	2,250	500
12	4½	4,000	175	3,840	210	3,680	245	3,520	280	3,360	315	3,200	350
12	5	5,500	150	5,200	180	4,900	210	4,600	240	4,300	270	4,000	300
12	5½	7,000	125	6,600	150	6,200	175	5,800	200	5,400	225	5,000	250
12	6	7,900	100	7,400	120	6,900	140	6,400	160	5,900	180	5,400	200
12	6½	9,000	85	8,600	102	8,200	119	7,800	136	7,400	153	7,000	170
12	7	10,500	75	10,100	90	9,700	105	9,300	120	8,900	135	8,500	150
12	7½	12,000	62	11,600	75	11,200	88	10,800	100	10,400	112	10,000	124
12	8	13,700	55	13,200	66	12,700	77	12,200	88	11,700	99	11,200	110
12	9	17,500	50	16,600	60	15,700	70	14,800	80	13,900	90	13,000	100
12	10	21,500	35	20,600	42	19,700	49	18,800	56	17,900	63	17,000	70
12	11	26,000	30	25,000	36	24,000	42	23,000	48	22,000	54	21,000	60
12	12	31,000	25	29,800	30	28,600	35	27,400	40	26,200	45	25,000	50
12	15	53,000	16	51,800	19	50,600	22	49,400	25	48,200	29	47,000	32
15	4	3,500	352	3,340	423	3,180	493	3,020	564	2,860	634	2,700	704
15	4½	5,000	275	4,800	330	4,600	385	4,400	440	4,200	495	4,000	550
15	5	6,000	225	5,800	270	5,600	315	5,400	360	5,200	405	5,000	450
15	5½	7,400	190	6,900	228	6,400	266	5,900	304	5,400	342	5,900	380
15	6	8,700	150	8,300	180	7,900	210	7,500	240	7,100	270	6,700	300

TABLE V.—*continued.*

Diameter of steam-cyl.	Diameter of pump	Steam 25 lbs.		Steam 30 lbs.		Steam 35 lbs.		Steam 40 lbs.		Steam 45 lbs.		Steam 50 lbs.	
		Gallons per hour	Vertical height in feet	Gallons per hour	Vertical height in feet	Gallons per hour	Vertical height in feet	Gallons per hour	Vertical height in feet	Gallons per hour	Vertical height in feet	Gallons per hour	Vertical height in feet
15	6½	10,000	125	9,500	150	9,000	175	8,500	200	8,000	225	7,500	250
15	7	12,000	115	11,500	138	11,000	161	10,500	184	10,000	207	9,500	230
15	7½	13,700	100	13,150	120	12,600	140	12,050	160	11,500	180	10,950	200
15	8	15,500	87	14,900	105	14,300	122	13,700	140	13,100	158	12,500	174
15	9	19,800	70	19,000	84	18,200	98	17,400	112	16,600	126	15,800	140
15	10	24,500	57	23,500	69	22,500	80	21,500	92	20,500	103	19,500	114
15	11	29,000	45	28,000	54	27,000	63	26,000	72	25,000	81	24,000	90
15	12	35,000	40	33,600	48	32,200	56	30,800	64	29,400	72	28,000	80
15	15	55,000	25	52,800	30	50,600	35	48,400	40	46,200	45	44,000	50
18	4	4,000	500	3,840	600	3,680	700	3,520	800	3,360	900	3,200	1,000
18	4½	5,500	400	5,200	480	4,900	560	4,600	640	4,300	720	4,000	800
18	5	6,500	325	6,200	390	5,900	455	5,600	520	5,300	585	5,000	650
18	5½	8,000	250	7,600	300	7,200	350	6,800	400	6,400	450	6,000	500
18	6	9,500	225	9,000	270	8,500	315	8,000	360	7,500	405	7,000	450
18	6½	11,000	200	10,500	240	10,000	280	9,500	320	9,000	360	8,500	400
18	7	13,000	165	12,300	198	11,600	231	10,900	264	10,200	297	9,500	330
18	7½	15,000	145	14,200	174	13,400	203	12,600	232	11,800	261	11,000	290
18	8	17,000	125	16,200	150	15,400	175	14,600	200	13,800	225	13,000	250
18	9	21,000	100	20,000	120	19,000	140	18,000	160	17,000	180	16,000	200
18	10	27,000	80	25,600	96	24,200	112	22,800	128	21,400	144	20,000	160
18	11	33,000	65	31,300	78	29,600	91	27,900	104	26,200	117	24,500	130
18	12	39,000	55	37,000	66	35,000	77	33,000	88	31,000	99	29,000	110
18	15	60,000	35	57,000	42	54,000	49	51,000	56	48,000	63	45,000	70
21	4½	5,500	525	5,200	630	4,900	735	4,600	840	4,300	945	4,000	1,050
21	5	6,500	450	6,200	540	5,900	630	5,600	720	5,300	810	5,000	900
21	5½	8,000	375	7,600	450	7,200	525	6,800	600	6,400	675	6,000	750
21	6	9,500	300	9,000	360	8,500	420	8,000	480	7,500	540	7,000	600
21	6½	11,000	250	10,500	300	10,000	350	9,500	400	9,000	450	8,500	500
21	7	13,000	225	12,300	270	11,600	315	10,900	360	10,200	405	9,500	450
21	7½	15,000	200	14,200	240	13,400	280	12,600	320	11,800	360	11,000	400
21	8	17,000	170	16,200	204	15,400	238	14,600	272	13,800	306	13,000	340
21	8½	19,000	150	18,000	180	17,000	210	16,000	240	15,000	270	14,000	300
21	9	21,000	135	20,000	162	19,000	189	18,000	216	17,000	243	16,000	270
21	10	27,000	110	25,600	132	24,200	154	22,800	176	21,400	198	20,000	220
21	11	33,000	90	31,300	108	29,600	126	27,900	144	26,200	162	24,500	180
21	12	39,000	75	37,000	90	35,000	105	33,000	120	31,000	135	29,000	150
21	15	60,000	50	57,000	60	54,000	70	51,000	80	48,000	90	45,000	100
21	18	89,000	34	86,000	41	83,000	47	80,000	54	77,000	61	74,000	68
26	5	6,500	675	6,200	810	5,900	945	5,600	1,080	5,000	1,215	5,000	1,350
26	5½	8,000	550	7,600	660	7,200	770	6,800	880	6,400	990	6,000	1,100
26	6	9,500	450	9,000	540	8,500	630	8,000	720	7,500	810	7,000	900
26	6½	11,000	400	10,500	480	10,000	560	9,500	640	9,000	720	8,500	800
26	7	13,000	350	12,300	420	11,600	490	10,900	560	10,200	630	9,500	700
26	7½	15,000	300	14,200	360	13,400	420	12,600	480	11,800	540	11,000	600
26	8	17,000	250	16,200	300	15,400	350	14,600	400	13,800	450	13,000	500
26	8½	19,000	230	18,000	276	17,000	322	16,000	368	15,000	414	14,000	460
26	9	21,000	210	20,000	252	19,000	294	18,000	336	17,000	378	16,000	420

TABLE V.—*continued.*

Diameter of steam-cyl.	Diameter of pump	Steam 25 lbs.		Steam 30 lbs.		Steam 35 lbs.		Steam 40 lbs.		Steam 45 lbs.		Steam 50 lbs.	
		Gallons per hour	Vertical height in feet	Gallons per hour	Vertical height in feet	Gallons per hour	Vertical height in feet	Gallons per hour	Vertical height in feet	Gallons per hour	Vertical height in feet	Gallons per hour	Vertical height in feet
26	10	27,000	170	25,600	204	24,200	238	22,800	272	21,400	306	20,000	340
26	11	33,000	140	31,300	168	29,600	196	27,900	224	26,200	252	24,500	280
26	12	39,000	125	37,000	150	35,000	175	33,000	200	31,000	225	29,900	250
26	15	60,000	75	57,000	90	54,000	105	51,000	120	48,000	135	45,000	150
30	6	9,500	625	9,000	750	8,500	875	8,000	1,000	7,500	1,125	7,000	1,250
30	6½	11,000	525	10,500	630	10,000	735	9,500	840	9,000	945	8,500	1,050
30	7	13,000	475	12,300	570	11,600	665	10,900	760	10,200	855	9,500	950
30	7½	15,000	400	14,200	480	13,400	560	12,600	640	11,800	720	11,000	800
30	8	17,000	350	16,200	420	15,400	490	14,600	560	13,800	630	13,000	700
30	8½	19,000	310	18,000	372	17,000	434	16,000	496	15,000	558	14,000	620
30	9	21,000	275	20,000	330	19,000	385	18,000	440	17,000	495	16,000	550
30	10	27,000	225	25,600	270	24,200	315	22,800	360	21,400	405	20,000	450
30	11	33,000	185	31,300	222	29,600	259	27,900	296	26,200	333	24,500	370
30	12	39,000	155	37,000	186	35,000	217	33,000	248	31,000	279	29,000	310
30	15	60,000	100	57,000	120	54,000	140	51 000	160	48,000	180	45,000	200

INDEX.

The black type figures indicate articles; the ordinary type incidental references The names of places where engines are at work will be of service to those who may wish to examine them, and make enquiries as to their merits in practice.

Printed in the United States
By Bookmasters